Science Observed

By the same author

University Perspectives (co-editor)
Wealth from Knowledge (co-author)
What Kinds of Graduates do we Need? (co-editor)
The Biochemical Approach to Life
The Teaching of Science

Science Observed

Science as a social and intellectual activity

F. R. JEVONS

M.A., Ph.D., D.Sc.
Professor of Liberal Studies in Science
University of Manchester

Distributed in the United States by
CRANE, RUSSAK & COMPANY, INC.
347 Madison Avenue
New York, New York 10017

LONDON · GEORGE ALLEN & UNWIN LTD
Ruskin House   Museum Street

First published in 1973

ISBN 0 04 502001 9 hardback
     0 04 502002 7 paperback

Printed in Great Britain
in 11pt Times Roman type
by Alden & Mowbray Ltd
at the Alden Press, Oxford

Preface

This book is about science, but not in the usual way. It does not set out to describe a part of the content of science, the facts and theories about the natural world which form its traditional subject-matter. Instead, it considers various ways of thinking about science – what kind of activity it is, how it works and how it relates to other activities. Now that it has itself become a major facet of the life of modern communities, how does it affect other social activities, and how is it affected by them? How fast has it in fact been growing? What is the nature of the approach that it adopts, and how does this compare with other forms of intellectual endeavour? How is the structure of scientific knowledge reflected in the organisation of science as a social system and of education in science? What motivates scientists, what goals do they aim at and what rewards do they value? Through what mechanisms does it yield wealth or other practical benefits? It is to questions like these that the book sets out to provide an introduction. They are relevant to every scientist and technologist, and to many others besides; but it is still comparatively rare for them to be included in educational programmes.

I have addressed myself to those who start by knowing little or nothing about the material and who do not want in the first instance to spend a great deal of time finding out about it. With this in mind, I have kept things simple. Had I been writing for a different readership, I would have adopted a different approach at many points.

Much of what is written about 'science and society' is intensely topical, dealing with issues of the day in which technical factors play a part. Here, although I have not avoided topical material altogether, I have concentrated mainly on what I take to be less ephemeral. I have tried to give a conceptual framework of some measure of permanence amid the year-to-year changes in issues raised by particular incidents. At the end of the book, the appendixes provide a set of readings, key passages selected with conciseness as one of the principal criteria. The main text provides the framework within which these are set, background explanations where they seem necessary, and discussion – often critical – of the views and conclusions set out by the authors of the selected passages.

It also does something to unify the different points of view from which science is considered. One underlying theme is that, although it is of course true that science and technology are highly specialised, they are not as totally different from other human activities as many people still believe. Neither as intellectual nor as social phenomena are they as sharply set apart as has sometimes been suggested in the past, and it is important that this should become more widely recognised so as to reduce the degree of isolation in which they find themselves. Another theme which recurs is the basic distinction between internal and external factors in the development of science. This dichotomy, and the conflicts which sometimes result, are worried at a number of times from different angles. The distinction originated in the history of science; philosophical and sociological considerations throw light on how the tension comes about; and it is a central issue in policy questions regarding scientific manpower and the direction of research effort.

The plan of the book has arisen out of the teaching I have done in this area. As far as possible, I have tried to base this teaching on critical discussion of reading matter. The reading matter should be carefully selected, realistically limited in extent and readily accessible to students so as to give them a common basis for fruitful dialogue in which the teacher, partly released from the need to impart information, can concentrate on putting questions and guiding discussion. Teaching of this kind is particularly valuable for those who have devoted their studies largely to science, where opportunities do not arise so readily to dissect open-ended issues and toss about the pros and cons.

To the students with whom I have had discussions I owe a great deal – almost as much as I do to my colleagues on the staff of the Department of Liberal Studies in Science. For specific comments I thank J. Ronayne, R. D. Johnston and M. Gibbons.

University of Manchester   F. R. Jevons
September 1972

Contents

1 Science and Society

The social relations of science

Many people are already concerned about the relations between science and the rest of society; more are becoming interested, and even more should be. Science and technology students with social consciences want to discuss the social responsibilities that go with the practice of their specialities. Some people urge that science and technology now have effects so powerful and all-pervasive that *all* students of these subjects should be *forced* to develop their social consciences by being made to discuss such issues. Students of non-scientific subjects often feel that, even if the goings-on inside laboratories must remain mysteries for ever hidden from them, they should at least make themselves aware of the social consequences. In society at large, plenty of people, both with and without technical training, take an active interest. To all these, this book is addressed.

We need first to formulate the problem in broad terms – to decide what kind of issues ought to spring to mind at the mention of phrases like 'science and society' or 'the social relations of science'. As a launching pad it is convenient to take the views of J. D. Bernal (1901–71). His book *The Social Function of Science*, first published in 1939, is recognised as a landmark in thought in this area. Himself a crystallographer of distinction, he took a long look, idealistic maybe but notably comprehensive, at the role of science in society: 'What science is – what science could do', as the sub-title of his book puts it. His left-wing inclinations are obvious enough. He was one of the small number of Marxist thinkers of any note that Britain has produced. Although this political background is important for understanding and evaluating what he has to say, it is often difficult to tell where its relevance ends; many aspects of the theses propounded with vigour from the Left find wide acceptance all the way across the political spectrum. In any case, there are plenty of people who are anything but Marxist in their own political sympathies but nevertheless acknowledge an intellectual debt to Bernal.

In a collection of essays published in 1964 to mark the quarter-centenary of the publication of *The Social Function of Science*, Bernal himself looks back on the main themes of the book. The essay in question, 'After Twenty-five Years', is reprinted in part in Appendix 1. Many of the chief issues are raised in it.

What are the main ways in which science and society relate to each other? How should we set out the main dimensions of the problem we have to discuss? The most common answers to these questions are likely, perhaps, to mention the waging of war, the pollution of the environment and the creation of wealth. Let us take up these topics in turn.

War
Understandably, in view of the period at which he was at the height of his impressive powers, Bernal was much exercised about the morality of the use of science in war. The military value of science was not new, of course. It was well enough recognised in the seventeenth century for Galileo, seeking patronage from the Duke of Florence in 1610, to take good care to mention the value of his mathematical mechanics for 'the practice of fortification, ordnance, assaults, sieges, estimation of distances, artillery matters, the uses of various instruments, and so on'.[1] But the scale was something quite new in the Second World War, and rather suddenly so, as is well brought out by the contrast with the First World War revealed in a story told by Conant.[2] Anxious to help with the war effort, the American Chemical Society contacted the Secretary for War to ask what they could do. The reply expressed polite appreciation, but thank you very much, the War Department already had *a* chemist. A mere quarter of a century later, the participation of scientists was much greater. Radar, the proximity fuse and above all the atomic bomb made the Second World War very much a matter for the 'back room boys'.

Bernal takes an idealistic line. 'The scientist as citizen is not in the first place a scientist, only in the second,' he writes, and calls for a 'collective effort to block out at least ideal policies which would have the general direction of making science serve the preservation and not the destruction of humanity. The more scientific effort that is directed to military ends, the more resistance it will create in the minds of scientists.'

Splendid sentiments, with which no one can fail to sympathise. But what if a scientist finds himself in a position where war seems to him, as a citizen, to be justified? Bernal himself must have been well aware of the conflicts of ideals that are bound to arise. In the Second World War, he at first 'solved the problems of conscience with ingenuity, by becoming the national (and the world) authority on the receiving end of bombing. By any conceivable standard, it was good to prevent people being killed by bombs'. Later, though, 'through ingenious steps he made his way via active bombing to Combined Operations, where he became scientific adviser to Lord Mountbatten', and he played an important part in preparing for the Normandy landings.[3]

Some countries devote substantial proportions of their research and development (R and D) efforts to military purposes. It is fashionable and quite easy to attack the distortion of research efforts by military pressures, but many of the people who do so find themselves unable to make reasonable guesses even at the order of magnitude of the sums involved, so it may be useful to give a few figures here, derived from the official published statistics.[4] In the United Kingdom in 1967–8, out of a total of £962 million spent on R and D, £226 million came from Government defence departments. Large though this figure may well seem, it does represent, at 24 per cent, a considerably smaller proportion of the total than that for 1961–2, which was 37 per cent (see also p. 101).

All right-thinking citizens naturally recoil with horror at the thought that war might be considered to be a good thing, but there is an uncomfortably well-supported view according to which it gives a powerful stimulus to science. The case is brilliantly put in *Report from Iron Mountain*, which purports to be the secret report of a high level Special Study Group to the United States Government. It argues that a state of permanent peace – defined not just as absence of war but as absence of threat of war – would, far from leading to Utopia, have the most dire consequences. The economy would be prone to depressions, social cohesion and political stability would be lost and, in particular, cultural and scientific activity would be deprived of their major driving force. 'War is the principal motivational force for the development of science at every level, from the abstractly conceptual to the narrowly technological. Modern society places a high value on "pure" science, but it is historically inescapable that all the significant discoveries that have been made about the natural world have been inspired by the real or imaginary military necessities of their epochs.' From the development of iron and steel to synthetic polymers and space capsules, military requirements have acted as powerful stimuli. 'The most direct relationship can be found in medical technology. . . . The Vietnam war alone has led to spectacular improvements in amputation procedures, blood-handling techniques and surgical logistics. It has stimulated new large-scale research on malaria and other tropical parasite diseases; it is hard to estimate how long this work would otherwise have been delayed, despite its enormous non-military importance to nearly half the world's population.'[5]

The anonymous author of *Report from Iron Mountain* may have been out to provoke and shock, but his arguments have too much foundation in fact to be brushed aside as merely perverse. If we want peace, will we have to sacrifice a lot of progress for it? Or can substitutes for war be found which provide stimuli of the necessary urgency without the accompanying suffering and misery?

Pollution

It may come as a surprise to some that pollution, an issue which today springs so very readily to the lips, does not figure prominently in Bernal: a salutary reminder that it is only recently that public opinion has got up so much steam about it. Quite unlike the question of war, it is only since the late 1960s that 'environment' has become a vogue, with legitimate concern liberally mixed with uncritical projection of trends, scaremongering and political cant to make a thriving industry of prophesying doom. The problems themselves are not so new. Smoke and other combustion products in the air, sewage and industrial effluents in rivers, lakes and seas, noise from machines for work and for travel have been features of advanced societies for generations.

What concerns us particularly here is the part played by science and technology in all this. To some it seems enough to point a finger at the obvious culprit and denounce as the cause of the trouble the relentless gallop of runaway technology. 'Stop the bandwagon, we want to get off', goes the cry. But to leave the diagnosis at that stage, satisfied with having found a scapegoat to be ritually slaughtered, is just an evasion of the real issues. It will do only for those who have made up their minds and do not want to be confused with facts. What the facts show is that science and technology are by no means the only sources of pollution; they do, on the other hand, provide means of control as well as contributing to the causes of the problems. In virtually every case, technical means are already available for avoiding or eliminating the unpleasantnesses. Most of the ground-level smoke in Britain comes from domestic fires.[6] How much does the open coal fire owe to science? Less than the smokeless fuels which could be used instead. How science-based is the fouling of beaches by oil? Less than the chemical dispersants used to wash it off.

What stands in the way of cleaning up the atmosphere and the waterways is not usually technical impossibility but lack of the will to commit the necessary resources. If cost were no object, it would be technically feasible to bring even the most heavily polluted river within a short space of time up to the standard required to support trout – but each one of those trout might work out exorbitantly expensive! And who is to pay? One of the features of pollution is that the effects fall largely not on the polluters themselves but on others. The sewage or factory waste that is disposed of so easily and cheaply by dumping in a river leads to increased costs or loss of amenity for others down river. It is possible to 'internalise the external costs' – to make the costs fall on the polluter himself – by legal and administrative means: he can be forbidden by law to discharge his rubbish or made to pay for the amount that he discharges. The prospect of fines or fees then exerts pressure on him to find other ways of disposing of his waste, or ways to avoid producing it.

But how high a standard is it reasonable to set? Costs usually rise very sharply as emission standards become more severe; most of the pollutant can be removed relatively cheaply, but to get rid of the last traces is exorbitantly expensive. Decisions have to be taken about how far it is worth going in each particular case. Different people and interest groups are bound to differ in their judgements as to what is justifiable. How much muck can be tolerated for so much money? What if the cost is not just lower profits or higher prices but loss of jobs because production becomes uneconomic? To clean up the River Don in Sheffield would be by no means beyond our technical capabilities, but unless specialist steel makers elsewhere were put under similar constraints, production in Sheffield would become uncompetitive.

Consider 'people pollution' – if that expression is not in itself a provocative contradiction in terms (what is an environment for if it is not for people?). Science has helped to bring about the decline in mortality which has resulted in the population explosion. It has also provided contraceptive techniques which could contain it. Whether and how much to use those techniques, however, is a question very much bigger than just technical feasibility. There are many other factors to be taken into account. Moral and religious scruples are strongly held by large groups of people. Some developing African countries believe that exhortations to them to encourage birth control are thinly veiled attempts to stop them rising in the pecking order among nations.

What does all this thorny complexity mean for scientists and technologists? Surely, that they should recognise their subjects as important but not exclusive or even dominant ingredients in a whole range of social problems. Science is not as special, not as different, not so much a thing apart from other activities as is sometimes made out. Societies have to set their priorities, and in the intricate social and political processes by which this is done scientists can and should play their due part. But the question remains wide open as to how much special responsibility rests on their shoulders by virtue of the technical expertise they command. Is it really any greater than that of other professionals who also have specialised knowledge and skills that they can bring to bear, such as economists, lawyers or accountants?

On political issues, by definition (p. 50), there is no unanimity of viewpoint. How far can scientists speak with one voice on the many matters of opinion and judgement that have to be decided? Balances have to be struck between the interests of different groups: oil tankers as against the seaside resorts and holidaymakers who want oil-free beaches; farmers as against anglers and others who want waterways unpolluted by organic matter; those who want cheap cars as against those who want air free of exhaust fumes. That is the human predicament: we cannot do without each other, but we are treading on each others' toes all the time; and scientists are likely to be just as divided among themselves on such matters as is society at large.

Faced with agonising choices, there may be a temptation to retreat behind the supposed ethical neutrality of science. Science and technology enlarge the bounds of what is possible;[7] society decides which of the many possible things to do. Why shouldn't scientists and technologists concentrate on the enlarging and let others do the deciding?

Not often is this possibility likely to turn out to be anything other than a comforting illusion. Responsibility cannot usually be divided so neatly. No more than anyone else can a scientist stand aside from – still less above – the hubbub and the turmoil of social controversies. As Bernal put it in the military context, the scientist as citizen is in the first place a citizen and only in the second a scientist. As a citizen he can no more expect that others must agree with him about the proper use of science than he can expect that they must vote the same way in elections.

For this reason, it seems unlikely that a 'Hippocratic oath for scientists' would get us very far. Every now and again there comes a call for such an oath, in which scientists should vow to work for some set of humanitarian ideals described in suitably vague terms. The difficulty is to decide, in any particular case, what course of action will actually bring us closer to realising the ideals. Usually, the call for a Hippocratic oath arises from the realisation that certain scientists are working to achieve some purpose of which some other group disapproves. No oath, however, can eliminate honest differences of opinion or get around the awkward truth that simple idealism often turns out to be inadequate in the face of the complexities of the real world. Sincerity of purpose is no consolation if the actual effects are other than those intended; it is sometimes the do-gooders who do most harm.

Particular cases abound which illustrate the complexities which can arise and drive home the point that scientists can neither abdicate responsibility to others nor arrogate to themselves the right to make ultimate decisions. Here I will take two examples. One comes from a nineteenth-century play and shows that, although popular concern is recent, the general issue is relatively old. The other is more topical: the use of DDT, which is very much a current problem.

In Ibsen's play *An Enemy of the People* (1882), the immediate issue is water pollution.[8] Dr Stockmann, medical officer of the Municipal Baths, discovers that the water supply is polluted by tanneries upstream. Being politically naive – extraordinarily so, even allowing for dramatic licence – he at first believes that he will be acclaimed as a public benefactor for this discovery. He is soon disillusioned. The only people who show any signs of gratitude are those who smell an opportunity for fomenting political revolution. There is inevitably some play with the notion that the pollution of the water is a symbol of the corruption of society. Officialdom, in the shape of the Mayor, points out to the doctor that disclosure of his findings will ruin the community. With its new Baths, the town has just established itself as a popular watering-place. Thus 'the matter in hand is not simply a scientific one. It is a complicated matter, and has its economic as well as its technical side'.[9] Emotional talk by the doctor about a 'permanent supply of poison' and 'making our living by retailing filth and corruption' will put off the summer visitors, and what will that do to trade and employment? So the Mayor wants to play the issue down and to suggest that the danger to health has not been conclusively demonstrated; in due course, modest improvements might then be made. Dr Stockmann, outraged by this attempt to muzzle him, calls a mass meeting; but public opinion is confused, ill-informed and easily swayed. As soon as they realise the threat to their prosperity and jobs, the townspeople turn against him and brand him 'an enemy of the people'. Inevitably he ends as an outcast, forced to the cynical political conclusion that the majority are *never* right.[10]

The passage of a century has not altered beyond recognition the constellation of factors and pressures operating in the situation Ibsen created. Some of the parallels with contemporary real life are quite obvious. In addition, there is one aspect which, although perhaps not intended by Ibsen, does add an extra touch of piquancy and realism. Dr Stockmann's evidence does not seem to be as conclusive as it might be. We are told only that there were some strange cases of illness among last year's visitors and that analysis of the water by a chemist at the

University showed decomposing organic matter. That does not add up to proof of typhoid infection via the water. We are not told of illness among the residents, and the visitors *could* have been infected before they came. A useful reminder that even the scientific aspects of social problems are rarely black and white!

A similar caution applies to the controversy surrounding the use of DDT, which is an outstanding instance of the many cases in today's world where it is by no means obvious which is the best line to take. DDT has become 'the most emotive and controversial of all pollutants; since the Test Ban Treaty it has surpassed even radioactive fallout as a matter for popular concern'.[11] It has been very successful in controlling the insects which spread some major diseases, and during the 1950s and 1960s freed some thousand million people from the risk of malaria. By keeping down crop pests, it substantially increases food yields. But it is notably persistent and has spread to every corner of the earth; it has been detected in rain falling on Shetland, in Arctic polar bears and Antarctic penguins. It concentrates along food chains and can reach high concentrations in the body fat of some fish and birds. Should its manufacture and use be stopped forthwith? Other chemical pesticides cost several times as much, and alternative methods of pest control are not yet adequately developed. The available evidence casts doubt on the harmfulness of DDT to man, as distinct from fish and birds. It is easy enough for those who live in well-fed European and North American countries to warn of ecological dangers and demand withdrawal; but what about those who live in countries where reinfection by malaria is a real hazard and where food supplies are precarious at best?

Wealth

That science does create wealth is widely accepted as self-evident. Some people seem to think that such use of science, although not as actively wicked as military applications, is somehow less noble than the pursuit of science 'for its own sake'. Such feelings are sometimes tinged with antipathy to the profit motive, expressed – as might be expected – by Bernal. However, the real dilemma for the scientist is not a matter of left-wing versus right-wing politics but of science for application versus science for knowledge. Whether the wealth that results from application is private, corporate or social is a secondary matter from the point of view of the research worker. What matters most immediately to him is whether the goals of his work are directed towards achieving aims external to science in the wider society or whether they are set internally within the community of scientists interested in advancing their subject. This external–internal distinction is an important one which is relevant in a number of contexts; in particular, it will need to be elaborated in discussing the motivation of scientists (Chapters 5 and 6).

The problem of wealth arises in particularly stark form in developing countries because there is so little of it there. Blackett[12] draws attention to the order of magnitude difference between incomes per head in the industrialised, relatively rich countries of Western Europe and North America on the one hand and less advanced Asian and African countries on the other. The gap in scientific development is even greater, and Bernal seems to suggest that because of this 'there can be no question of an automatic and independent "catch-up" on the part of the developing countries'.

How firmly based, though, is the assumption that science really does lead to wealth? Perhaps it is necessary at this stage to distinguish between science and technology rather than lumping the two together. Popular usage often neglects to make the distinction and 'science' is used as an umbrella term to cover both (although, as engineers sourly complain, the moon shot that succeeds tends to be hailed as a 'scientific triumph' whereas the one that goes wrong is merely an 'engineering failure'). However difficult it may be to place borderline cases, there clearly are important differences between science and technology in their typical forms. A moment's thought about the real basis for the distinction suggests that whereas science aims at knowledge and understanding, technology is the ability to do things. In economic terms, science *costs* money; it is only technology which may – with skill and luck – be made to pay. Engineering products form an output which can be sold; but the search for knowledge about subatomic particles or molecules or living cells gives nothing of direct money value to set against the input of effort and materials.

It is believed by many, of course, that technology is built on the application of scientific discoveries. This belief finds expression in the view that in modern times application has been following ever more rapidly on the heels of discovery. Bernal himself states this view clearly. 'The time lag of the application of research has greatly shortened; new ideas can come into application, especially in fields which are advancing most rapidly, like those of control mechanism, within a year or two of their first discovery.'

Most people can pick out of their fund of general knowledge at least one or two examples against which to test the assertion about a historical trend towards shorter time lags. Plausible though it seems at first, the harder one looks at particular cases, the less clear the situation becomes. Atomic energy is an example that springs readily to mind. Clearly it arose from the discoveries of nuclear physics in the first half of the twentieth century. But which is the proper starting point to take? The discovery of nuclear fission in 1938; or the discovery of the neutron by Chadwick in 1932; or Einstein's equation $e = mc^2$ which predicted the convertibility of mass into energy; or something in the centuries-old tradition of physics which led up to Einstein? As another example, take the case of radio. Fleming's invention of the thermionic valve, Thompson's discovery of the electron, Hertz's detection of radio waves, Maxwell's electromagnetic theory and Faraday's discovery of induction all seem reasonable 'starting points', but they were spread over a period of three-quarters of a century.[13] If scientific ancestries are so extended over time, the choice of starting points, and hence the calculation of time lags, must necessarily be to some extent arbitrary.

Besides, there are many important technologies which can hardly be said to be based on particular scientific discoveries. What about baking, brewing, building and the manufacture of metals and of glass? All of these have been profoundly affected by science, but in their origins they are based not on science but on crafts.

There seems to be more complexity than is often thought in the relations between science, technology and wealth. We will need to go into the matter more fully later (Chapter 6).

How far does science extend?
When Bernal writes that 'the scientist as citizen is not in the first place a scientist, only in the second', he raises the issue of how far science extends. What are the proper boundaries of the subject-matter and the methods of science? Just those of the

broad areas of physics, chemistry and biology which by tradition are labelled 'science' in the educational system? Or is it proper to speak of 'social science' and 'behavioural science'? Can history be written 'scientifically' and is there a 'science of linguistics'?

In a revealing passage, Bernal makes his own view clear. 'Any human activity and any branch of that activity is a legitimate subject for scientific study, and subsequently for modification in the light of that study. Once this is accepted in practice, which implies the provision of research workers to carry out these studies, the way is open to a new level of man's control of his environment, one in which economic and social processes become scientific through and through.'

Bernal was writing against a background of wartime operational research, which had led not only to greater understanding in detail of individual operations but also to better integration of combined operations by land, sea and air. Applied to the peacetime economy, this kind of approach opens up a huge new vista: the possibility of a science *of* society. Here phrases like the 'social function' or the 'social relations' of science take on a new and much wider meaning. A science *of* society implies much more than the impact *on* society of the products of science and technology. It includes not just the ways in which new chemicals and new machines have affected daily life but also encompasses dreams of a society in which social affairs are permeated by the methods and ways of thought of science and in which there is scientific study and integrated control not just of isolated sectors and aspects of society but of society as a whole.

Such dreams are characteristically Marxist (not uniquely so – Marx was in any case not the first to put them forward – but characteristically in the sense that they have formed important elements in Marxist thought). Bernal feels sure he knows what kind of society would emerge if scientific principles were followed in constructing it. 'I feel confident that the ultimate pattern will, so to speak, impose itself the moment its logic is fully appreciated.' 'The scientific and computer age is necessarily a Socialist one.'

These are big issues and it is worth pondering over the implications. There is, of course, one big attraction about the possibility of arriving at decisions about society by logical means:

such decisions should command universal assent. Everyone would then agree about the best course of action instead of eternally arguing and haggling and lobbying for sectional interests. But many people feel that scientific managerialism has already gone too far, that we are already over-studied and over-manipulated by experts (cf. p. 48). On this issue the New Left is largely in accord with the traditional Right, the political spectrum having bent back on itself so that the two extremes nearly meet in one. Ultimately, Bernal envisages up to 20 per cent of the population in scientific functions in the wide sense – 'not necessarily scientific research and development only, but scientific production and scientific administration'. For some, this possibility of a scientifically run society may be a wishful dream, but for others it is a nightmare.

One other important issue is raised by Bernal's vision. How far is it really true that scientific method is a procedure for arriving by logical means at unique best solutions? A closer look at how natural scientists operate will show that the methodological prescription cannot be formulated quite as simply as that (Chapter 3).

The dilemma over planning

It is a matter of common knowledge that science has been growing rapidly; later (Chapter 2) we will look at how rapid the growth has been. Bernal recognises that the change in quantity has made it necessary to view science in a qualitatively different light. Having complained in 1939, in *The Social Function of Science*, of the gross inadequacy of the resources being devoted to science, he admits in 1964 that 'now there is a different situation – it is the large scale of expenditure on science rather than a small scale that must be considered'.

The new magnitude makes inescapable the problem of whether and how to plan science. Bernal naturally believes in planning. 'We need a strategy for research which must be based on a *science of science*.' 'The whole problem – economic, scientific, and political – must be regarded as one of a planned operation.'

It makes admirable sense, naturally, that planning should be comprehensive, that economic, scientific and political aims should be in harmony and not at cross purposes. As with all planning, however, there is a snag: what about the freedom of the individual? In this particular case, what about freedom of research? Here is a dilemma. We want good planning to make things run smoothly, but freedom is a powerful counter-attraction. Amongst the scientific community, the right to choose one's own problems and mode of attack is a cherished ideal. Even among the public at large, there are plenty of people who feel that a scientist should be as free to do what research he likes as a painter should be to paint what he likes. Much virtuous prose has been written about freedom of research. To work up a bit of passionate eloquence in its defence is as easy as it seems churlish to attack it.

Solid argument in its favour, as distinct from emotive verbiage, is more difficult to find. It is important, though, to look at the case seriously if we are to assess the implication of Bernal's call for planning. One of the most powerful short statements of the anti-planning position is given by the chemist and philosopher Michael Polanyi in his article 'The Republic of Science'.[14] Only by allowing free play to the individual initiatives of scientists, maintains Polanyi, can we optimise the use of scientific resources. His argument proceeds by analogy with the kind of economic liberalism that regards a free market mechanism as the way to optimise the use of economic resources. With each research worker free to make his own informed judgement as to where and how, in the current state of knowledge, he can best make a contribution, it is as though there were something like Adam Smith's 'invisible hand' to direct efforts into those areas where it is likely to have the highest scientific value. Co-ordination is achieved by each individual adjusting his own work to the results obtained by others. Any attempts to organise science under a central authority would paralyse this co-operation, reducing its effectiveness to that of the single central authority. Moreover, it is just not possible to guide science into socially beneficent channels. However generous the sentiments which give rise to the desire to do so, they do not make the impossible practicable. Science 'can advance only by essentially unpredictable steps, pursuing problems of its own, and the practical benefits of these advances will be incidental and hence doubly unpredictable'. Thus 'any attempt at guiding scientific research towards a purpose other than its own is an attempt to deflect it from the advancement of science'. It follows that the allocation of money to science should be guided by the advice of scientific and not of other opinion, and 'adulteration' of science by political, economic, commercial or other outside interests is to be avoided.

This adds up to a far from negligible case. What can be said against it? The alleged unpredictability of applications is a key point. How laudable is it for scientists to be concerned only with increasing knowledge, refusing to consider anything else? If that is what they do, it can hardly be more than coincidence if the kind of knowledge they gather is the kind that will make society better rather than worse. Scientific research 'pursuing problems of its own' acquires a momentum of its own. Institutionally separated in places like universities, it develops its own sets of values and rewards (Chapter 5). They encourage choice of projects according to standards internal to science, aiming to solve the mysteries of the constitution and working of the physical universe but neglecting society's problems of production, transportation, disease and so on. The knowledge that is generated may be of a kind that almost begs to be misused, like that concerning the energy of the atomic nucleus, or the structure of human genes, or drugs that affect behaviour. To shield behind unpredictability is a negative and unconstructive posture. Is it anything more than an excuse for not trying harder to predict? Does it not mean abnegating responsibility for the consequences of one's actions, akin to the schoolboy who says it is not his fault the window was broken because he certainly didn't mean to break it and anyway it was the football and not his foot that hit it?

Scientific choice: the Weinberg criteria
An apparently easy compromise offers itself as a way of resolving the dilemma over whether to plan: let us reconcile ourselves to the necessity of planning and controlling work from which applications can be foreseen but let science which is judged to be pure and without foreseeable applications remain free. In this way, both social morality and scientific honour could perhaps be salvaged. That is not the way things work out in practice, however. What remains free is what is cheap. Experiments that require little in the way of apparatus, materials and manpower can easily be done. It is when demands on resources become substantial that the real test comes for freedom of research. Even if nobody wants to direct and control

basic research in detail from the centre, is it feasible to do without central bodies which, by allocating grants and scrutinising expenditures, cannot help having a very decisive say as to what can and what cannot be done?

In the years immediately following the Second World War, when there was immense faith in the power of science, Bernal felt, like many others at that time, that scientists deserve to be given all they ask for. 'The principle first enunciated by Professor Blackett, that allocation of money to science should be made in the measure of what a competent scientist can usefully spend and not according to what he can just manage on, should be the basis for our post-war science.' It would be comfortable and pleasant in many ways if such open-handed largesse were still justifiable today. But is munificence with public money not stretched to the limit when nuclear physicists say that, to investigate the fundamental structure of matter, they need particle accelerators whose costs run into nine figures? With financial stakes of such magnitudes, 'grantsmanship' inevitably overlays disinterestedness. No one can still pretend otherwise in the face of Greenberg's impressively documented survey of the American scene. 'Science', concludes Greenberg, 'like agriculture, the military, labour, business, or the civil rights movement, has its vested interests, elites, downtrodden, alliances, bosses, loves and hates. The politics of science is in essence no different from other politics.'[15]

Nuclear physics is the current *cause célèbre* because it poses the problem in its most acute form. Here is good science, indubitably, giving us what is literally the most fundamental knowledge of the physical universe, but hugely expensive and offering no foreseeable practical applications. For science policy in Britain, it was a turning point when, in 1968, the government rejected a proposal to join an international venture to build a 300 GeV proton synchrotron at an estimated cost of £150 million, to be shared between several European countries. The decision was taken against the advice of both the Science Research Council, from whose budget the money would have come, and the Council for Scientific Policy, which took a general overview of all the research councils, and the affair caused something of a stir.[16] Although permission was later given for a somewhat less expensive proposal to achieve similar objectives, the rejection marked very clearly the growth of public recognition that basic science is no longer a sacred cow that must be fed whatever the cost. So many other things compete for public money and trained manpower. 'In a world plagued by misery, is it decent for fine minds and great wealth to be dedicated to the interior of the atom and the mysteries of the planets?'[17]

When resources are not limitless, choices have to be made between possible projects and lines of research. Choices always tend to be agonising, but the agony becomes milder if there are clear criteria by which to make them. There is one set of criteria which has been particularly influential in the debates over scientific choice which became prominent in the 1960s and which seem bound to become a permanent feature of the science policy scene. This is the set put forward in 1963 by Alvin Weinberg, Director of the Oak Ridge National Laboratory, Tennessee and a former member of the Science Advisory Committee of the President of the United States. A part of his paper is reprinted in Appendix 2. In it, he argues that the criteria should be of two kinds, internal and external. The internal ones, generated within the scientific field itself, concern the quality of work in it; the main issues here are whether the field is ripe for exploitation and whether the scientists active in it are of high quality. In addition to these – and in Weinberg's view more important – are external criteria, of which he identifies three. One is scientific: are advances in the field important for other related fields of science? The second asks whether the field has potential technological value, and the third whether it serves desirable social ends.

It is a salutary exercise for any scientist, or indeed any science student, to make the effort to see how well his own field of science measures up against the Weinberg criteria.[18] The distinction between internal and external criteria has come into wide use. It may help to avoid confusion to point out that most people use 'external' to mean external to science as a whole, whereas Weinberg, since he bases his formulation on a field of science, calls the relevance of one field to other fields of science 'external'.

The role of individuals
One final point arising out of Bernal's article merits consideration here. Again it is one which raises a big question mark. It concerns the role of individuals. To an increasing extent, they

cannot do scientific work alone and independently. Complex projects can only be undertaken by teams whose work is organised. Does this mean that the days of the virtuoso solo performer in research are drawing to a close? Can organisation, together with an increased understanding of the kinds of methods to be adopted, eliminate the need for individuals of genius?

Bernal seems to suggest that this might be so. The electronic computer, he says, 'was not an invention of any one person – it did not require genius, but simply application of known methods to known problems'. Scientific method 'is a method which in itself can be *counted on* to generate more and more of these great achievements and transformations'.

There is a strain here of the kind of historical determinism which is sometimes associated with Marxism. History will take its course whatever any of us do, it seems to be implied. Men of outstanding genius are not necessary for progress; at best they can, perhaps, speed up what would otherwise have taken a little longer. Cynics may object that this is the kind of thinking which lowers the valuation of the individual human being and can be used to justify or excuse political purges in totalitarian states. But the concept has its counterpart in non-Marxist thought. The American economist J. K. Galbraith, that shrewd analyst of the power of the large industrial corporation, suggests that 'the real accomplishment of modern science and technology consists in taking ordinary men, informing them narrowly and deeply and then, through appropriate organisation, arranging to have their knowledge combined with that of other specialised but equally ordinary men. This dispenses with the need for genius. The resulting performance, though less inspiring, is far more predictable'.[19]

Must we accept, then, that organisations run by organisation men can and will replace outstanding individuals? Perhaps the alternatives need not be put quite so bluntly. It could be that what we need for the future is a different kind of genius: one who shows his genius in working with others rather than alone. There is some evidence that this is so for technological innovation in industry. Outstanding individuals are often important, but they are less often the 'lone inventors' beloved of romantic tradition than 'product champions' with the ability to pilot their pet projects through the unromantic bureaucracies of the modern industrial system.[20]

Be that as it may, it is clear that a small select elite of scientists is not enough by itself to give modern industrial societies the benefits that science is capable of conferring. Bernal draws out the moral for education. 'There is a realisation, which is just beginning in some of the older industrial countries, that not only a small section of a professional class needs such education [i.e. scientific education], but that it must also be spread throughout the whole population.' Stating this egalitarian ideal does not, of course, solve the problem of what kind of science education it should be that is provided for everybody and of deciding to what level the science is to be taken. Few people would want all graduates to be science graduates, even in Bernal's wide interpretation of 'science' to include not only natural science but scientific production and administration. Nevertheless, the contention that science should form a vital part of everybody's education seems unexceptionable.[21] It is, surely, one of the least controversial aspects of Bernal's vision of a society permeated by science and utilising it to the full.

2 Trying to Measure Science

Possible approaches

To plan better for science, to ensure that its needs are met in the best possible way and that it does the maximum amount of good and the minimum harm, we need to know as much as possible about it. 'Knowing about science' in this context does not mean just knowing something of the facts and theories about nature which make up the context of science, though there must, of course, be people with this kind of knowledge. Rather, it means studying science as one might study, say, a living organism: beginning with a thorough description of its features, including its parts and their relations to each other; going on to investigate how it grows, under what conditions it thrives, how its metabolic machinery works, what its behaviour patterns are like, and so on. In short, we need the sort of knowledge of science that could underpin science policy, making it more effective by giving greater understanding, just as economic policy-making is strengthened by a sound grasp of economic principles.

The aim seems unexceptionable But how are we to set about achieving it? What kind of approaches would it be appropriate to adopt to get that kind of knowledge? It depends, of course, on what specific questions about science we want answers to. Once the questions have been formulated, the answers are half predetermined. For the moment, let us keep the discussion very broad, not concentrating on specific policy issues but thinking in general terms about how to get a better and deeper understanding of the characteristics and mode of operation of science.

One possible mode of approach is sociological. There is a sociology of science just as there is industrial sociology and sociology of education. One could study the organisation of scientific institutions and the formal structures and informal mechanisms by which decisions are taken; or communication patterns, estimating the frequencies with which scientists use various channels for giving and getting information; or the motivations of scientists, what attitudes they take, what value systems they adopt, what rewards they value. Typically such matters would be investigated by questionnaires and interviews. Some work of this kind is discussed later in this book (Chapter 5).

Another possible mode of approach is through philosophy of science. Philosophers – and philosophically inclined scientists – have long debated the nature of scientific method and the status and functions of scientific theories. Some of these issues are also raised later in this book (Chapters 3 and 4). It is a fact, though, that philosophy of science has never made much impact on practising scientists and most of it is totally ignored by nearly all of them. Although this is a pity in a way, scientists can be partly forgiven for feeling that they are not missing much and that they have better things to do. We will have to beware of getting into the kind of academic backwater where philosophers argue more against each other than about science.

A historical approach is a third possibility. Philosophy and history of science have traditionally been closely associated with each other. History must, of course, if it is taken to include recent as well as early history, be the main storehouse of facts against which to check philosophical views about science. But, if the history is to be more than a mere chronology of dates and events, it must involve interpretation by the historian. Is it then any more than the opinions of an individual scholar, illustrated by the facts he has himself selected from the storehouse? History can be written in many ways and take many forms. Merely to appeal to history still leaves open the question of what aspects of science we want it to illuminate for us.

What Bernal called for was a *science* of science (p. 25). The first science in this double-barrelled expression means that science is to be studied scientifically. What is a 'scientific' way of studying something? Customarily it means an objective and preferably quantitative approach. Although later (Chapter 3) we must undertake a thoroughgoing criticism of this view of what it means to be scientific, let us here accept it, if only because it corresponds to what many people in fact want of a science of science.

Can we, then, get some hard facts and figures about science –

something that goes beyond the subjective views of individual interpreters? In this field, outstanding contributions have been made by D. J. de Solla Price. Trained originally as a physicist, he turned to history and became a professional historian of science. Some of his main results are set out in his book *Little Science, Big Science*, in which he describes his aim as follows: 'My goal is not discussion of the content of science or even a humanistic analysis of its relations. Rather, I want to clarify these more usual approaches by treating separately all the scientific analyses that may be made of science. Why should we not turn the tools of science on science itself? Why not measure and generalise, make hypotheses, and derive conclusions?'[1] With this object in mind, he deals statistically with some general problems of the size and growth of science.

Exponential growth

One basic conclusion, extensively documented by Price, is that science grows exponentially. This fact has become common knowledge; it makes easy copy for newspapers and is freely bandied about in conversation. But what does it mean in more precise terms? Science has no mass or length; quantity of knowledge defies precise determination; so what is there about science that one can measure? A few moments' thought suggests three main possibilities. Two of them are inputs into science – the amount of money spent on it and the number of scientists active in it. The third is a measure of output – the number of publications that result.

It is Price's contention that, if any sufficiently large segment of science is measured in any reasonable way, it is found with impressive regularity that growth has been exponential. A useful way to characterise the rate of exponential growth is by the time needed for the magnitude to double. Taking manpower or publications as indicators of the growth of science, doubling times are found commonly to be around 10 to 20 years. For instance, Price[2] gives a doubling time of about 20 years for college entrants per 1,000 population and for important physicists; of about 15 years for B.A.s and B.Sc.s, for the numbers of chemical compounds known, of scientific journals and of scientific abstracts in all fields; and of about 10 years for the numbers of engineers in the United States and for the literature in fields like the theory of determinants, non-Euclidean geometry and experimental psychology.

To bring out the significance of such figures, take the doubling period of the number of scientists as 15 years and note that in each doubling period about as many new scientists come into being as in all previous time. (This is simple arithmetic: 8 is close to $4+2+1$, 16 is close to $8+4+2+1$, and so on.) Since the working life of a scientist is likely to be two or three times as long as the doubling time that has been assumed, most of the scientists who have ever lived are alive today. This is what Price refers to as the 'immediacy' of science. For a doubling period of 15 years and an average working life of 45 years, the 'coefficient of immediacy' is seven-eighths; that is, there are seven scientists alive today for every eight there have ever been.

Such a result may seem surprising at first sight. On second thoughts, it is in one way but not in another. Contrary to what seems sometimes to be assumed, there is nothing unusual about the fact that growth has been exponential. Exponential growth is merely growth at compound interest, which is quite a common way for things to grow. It just means that the amount by which it grows in any time interval is proportional to its total magnitude at that time. Thus it is no occasion for surprise that a sum of money invested at a fixed interest rate grows exponentially. So does the number of bacteria in a freely growing culture. So also has the total human population of the world grown. But – and this is the point really worth noting – for the total population the doubling period has been nearer 50 years than 15. We hear a lot about the population explosion, but the increase in the number of people in the world has been much slower than the increase in the number of scientists; so that, while most of the scientists who have ever lived are alive today, most of the people who have ever lived are dead.

Take another comparison to help to put this matter of growth rates into perspective. A politician who makes the promise – as has been known to happen! – to double his country's real income per head in 35 years is merely making political capital out of the arithmetical wonders of exponential growth. His promise amounts only to raising the standard of living by a very modest 2 per cent per year, as can be calculated quickly by anyone who knows that percentage growth per year is about 70 divided by the doubling period. A doubling time of 15 years, by contrast, corresponds to an annual increase of about 5 per cent. What is unusual about the growth of science, then, is not its exponential nature but the short doubling time.

Does the immediacy of science mean that the history of science in earlier centuries is negligible? There is 'more past to live in', so to speak, if one considers the history of politics or war. Science is mostly a matter of the last few decades, which is

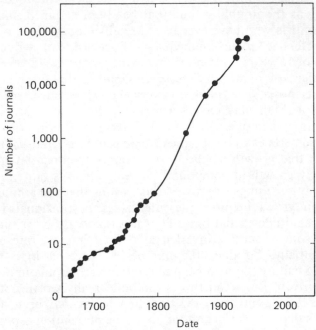

Fig. 1. Total number of scientific journals as a function of date. Numbers recorded are for journals founded rather than those surviving. They are for all periodicals containing any 'science' rather than for 'strictly scientific' journals. Tighter definitions might reduce the absolute numbers by an order of magnitude, but the general trend remains constant for all definitions.

Source: Price, *Little Science, Big Science*, p. 9.

not what is usually understood in the academic context as 'history of science'. But Price is a professional historian of science and he does not at all take the view that the earlier history of science can be ignored. On the contrary, he is struck by the constancy of the phenomenon of immediacy over the last 300 years or so – a constancy which only history could have revealed. A plot of the logarithm of the number of scientific journals against date, for instance, gives quite a good straight line, indicating exponential growth (Fig. 1). It extrapolates to zero around the end of the seventeenth century, the time of Newton, which it is reasonable on other grounds as well to regard as the take-off point for the sustained rise of modern

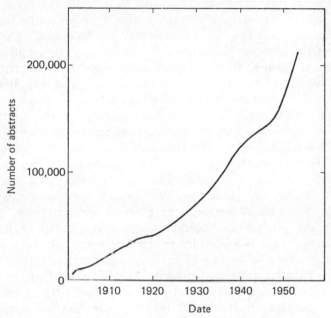

Fig. 2. Total number of *Physics Abstracts* published since 1900. Growth slowed at the times of the two World Wars, but continued exponentially after them.

Source: Price, *Little Science, Big Science*, p. 18.

science. A high degree of immediacy has therefore been characteristic of science for centuries; throughout this period it has always been true that most science is recent science.

Price takes a very uniformitarian view: growth has been proceeding steadily, not in violent bursts. There is a popular impression that it was the cataclysmic events of the Second World War which projected science into prominence. For Price, however, they produced only a temporary perturbation in the apparently inexorable course of exponential growth. Thus the curve for the number of physics abstracts (Fig. 2) shows merely a displacement to the right due to 'secrecy loss'.

Levelling off

Inexorable the exponential growth may have been in the past, but quite clearly it cannot continue to be so indefinitely. Extrapolation into the future leads quite soon to various absurdities, such as having more scientists than there are men, women and children in the population. It is unavoidable that the growth rate must level off, and Price calls this a second basic law in the analysis of science (the first being the law of exponential growth). In many cases of biological growth, such as the number of bacteria in a culture or the length of a beanstalk, the growth curve eventually becomes S-shaped; the slope stops becoming ever steeper, as in pure exponential growth, and flattens out to approach a saturation level.

Price suggests that the same is happening, or beginning to happen, to science, though of course the flattening need not be smooth but may well be marked by more or less violent fluctuations. Here is something which does seem new in modern as distinct from older science. The long period of exuberant growth looks like a sort of prolonged adolescence which cannot go on much longer; with a few more doublings, manpower and expenditure would hit the ultimate ceiling at 100 per cent of what is available. In the real world, slowing down must begin long before that. The transition from free exponential growth to more restricted growth introduces new strains and stresses and brings with it a need for adjustments which are not always painless. It may well be that here lies the soundest explanation at the macro level for all the current concern over problems of policy for science, of manpower, education, information, support of research and criteria for the allocation of resources – in short, for the fact that science has become self-conscious, as Bernal put it. Maturity and self-awareness bring the realisation that it is not possible to do everything, that choices have to be made, that not all change can be accommodated within expansion, that 'as well as' must sometimes become 'instead of'.

In Britain, consideration of growth rates and the ultimate inevitability of their decline made a conscious impact on science policy by 1966. In that year, the first report of the newly formed Council for Scientific Policy did not question that slowing of growth must take place; the matter at issue was only its timing and extent, 'when, and at what rate, and on what

criteria, the levelling-off of the growth rate should take place.'
Undoubtedly the Council was bearing in mind the rate of
increase in national expenditure on research and development, which rose between 1961-2 and 1967-8 from £658
million to £962 million. Growth was particularly rapid in the
universities and further education establishments, where
expenditure increased in those 6 years from £32 million to £75
million.[3]

The response of the scientific community to such restrictionist
thinking was far from enthusiastic.[4] One could hardly have
expected a warm welcome, of course, but there was some
justification for the fear that talk of S-shaped curves would be
taken up with undue enthusiasm by those in the government
machinery who are always – and especially soon after elections,
when the next election is not an immediate prospect – on the
look-out for ways to cut expenditure of public money. Cuts
involve agonising decisions and those who have to make them
are liable to snatch at any straw which looks like a rational
justification for pruning one thing rather than another.

How much does Price-type analysis of growth curves really
offer in the way of genuine guidance for science policy? Is it
a step towards the sort of 'science of science' for which Bernal
called as an aid to planning? It remains very much a matter for
debate what is the right saturation level towards which growth
should aim. There can be very different views about this, even
allowing for differences in the definition of 'science' (p. 24).
Besides, Price's treatment seems rather dispassionately 'physi-
calist', treating science by something analogous to the kinetic
theory of gases and hiding behind the statistical safety of large
numbers. Science is not really like a gas composed of molecules
in random motion. On the contrary, it has a highly differentiated
structure. Less global figures for smaller sectors of science over
smaller time spans do not often give curves as neat as the ones
that are gleefully published. Nor is it at all obvious that we
should want them to be. Do we not want to retain freedom to
run down or phase out some areas of research while vigorously
promoting others? That is what we have to do if we want to
make science help to shape the sort of society we want rather
than allow society to be shaped by any alleged 'inevitability'
of the course that science takes. Price's curves must not be
misinterpreted as historically determinist laws.

Productivity

One factor entirely left out from the analyses mentioned above is the quality of research. Obviously only some of the work that is done is good and much is indifferent at best. Are there any possible ways of measuring quality?

Clearly it is going to be difficult to find acceptable measures, if only because there are different standards by which to judge. One could use special awards like Nobel Prizes and membership of learned societies like the Royal Society to indicate scientists of real eminence. Price discusses[5] listings of scientists in a biographical compilation, *American Men of Science*. The numbers listed increase from 4,000 in 1903 to 96,000 in 1960 (omitting social sciences). The numbers 'starred' as being specially distinguished increase less rapidly, in line with other findings that the more selective indicators of growth have longer doubling times than those which do not distinguish high grade from low grade work. But editorial judgements of standards of scientific eminence must be at least partly subjective and they form a very dubious basis for elaborate numerological exercises.

One measure which is reasonably objective is the number of papers published by a person in scientific journals. Of course, this measure also is imperfect. One paper may report an epoch-making discovery while another describes only some relatively trivial redetermination of certain experimental values or variations in the conditions for a reaction. There is a difficulty also over multiple authorship, which is increasing as a proportion of all work. Because publications are used as a basis to help in deciding appointments and promotions, there may be a tendency to try to spin out a given amount of work into more papers than are really necessary. Whatever it is that is measured by a paper count is perhaps better described as productivity than as quality. But at least it is something measurable by which one can compare the output of different scientists. Though there are flagrant anomalies in individual cases, there does seem on the whole to be reasonably good correlation between productivity in terms of papers and eminence as judged in other ways.

Using paper counts, one can investigate the frequency distribution of scientific productivity. What kind of distribution would one expect? Would one guess that the productivity of most scientists clusters round some average number, with only exceptionally good or exceptionally bad ones producing many more or many fewer? In fact, one finds quite a different kind of distribution, as has been known since 1926, when A. J. Lotka noted that it approximates to an inverse square relationship: the number of people who produce n papers varies as $1/n^2$. There are many who produce only one paper, only a

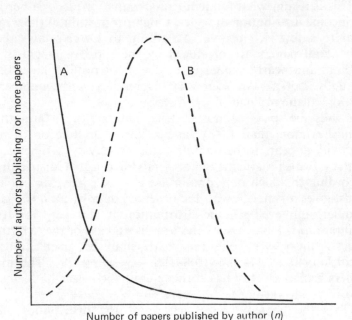

Fig. 3. Number of authors publishing at least n papers as a function of n. The observed distribution resembles curve A rather than curve B; the 'typical scientist' produces much less than the average number of papers per scientist.

quarter as many who produce two, one-ninth as many who produce three and so on. This relationship holds reasonably well[6] for sources as widely separated as the *Philosophical Transactions of the Royal Society of London* in the seventeenth and early eighteenth centuries and *Chemical Abstracts* in the period 1907–16. Working out the distribution cumulatively, it turns out that approximately one in n authors produces n or more papers. Only 20 per cent of authors produce five or more papers.

Thus the distribution of productivity is nothing like the sort of normal distribution that one gets for chance events. Plotted as a graph, the curve resembles A of Fig. 3, not the familiar bell-shape of B. The great majority of people who publish at all do not publish more than a very few papers. Many manage to produce one or two as research students but never any more. Only quite a small and select minority produce more than a few.

This is a somewhat 'undemocratic' situation; it is rather like an income distribution in which a high proportion of the wealth goes to a few rich people. According to Lotka's law, about half of all papers are produced by the top 6 per cent of all authors and nearly a quarter by the most prolific 1 per cent. Actually, Lotka's law somewhat overestimates the number of authors at the top end of the distribution. It is easy to see that this must be so. It cannot be true that one in 1,000 authors publishes more than 1,000 papers. The highest recorded productivity appears to be the 995 items by Cayley, a nineteenth-century British mathematician. A modified Lotka distribution, according to which only one in n^2 authors achieves n publications, gives a better fit with the observed data where n is a large number. But even so, the distribution is still very far from egalitarian. It looks, the cynics say, like a case of the 'Matthew effect': 'unto every one that hath shall be given' (Gospel according to St Matthew, chapter xxv, verse 29). The more papers a man already has to his credit, the easier he seems to find it to add further to his score.

Does the 'undemocratic' nature of the distribution have undemocratic implications for education? If such large proportionate contributions are made by relatively few people, it might seem best to concentrate on giving the best possible education to a small number instead of spreading resources more thinly in an attempt to achieve equality of educational opportunity. Is this a valid argument for an elitism which looks as though it might clash head-on with the call (p. 31) to spread education throughout the whole population?

Obviously no definitive answer can be given here to this highly controversial question, but there is one point which is worth noting in the present context. In most cases, in the experimental sciences at least, prolific authors can achieve high scores only because they have co-workers, students and assistants. Such apparently lesser mortals are also important. Solo performances in research are becoming relatively more rare. The trend in indicated in the increasing 'multiplicity' of authorship. In *Chemical Abstracts*, single-author papers had declined to account for less than half the total by 1960, while papers with four or more authors are increasing more rapidly than papers with smaller numbers of authors (Fig. 4). Thus there

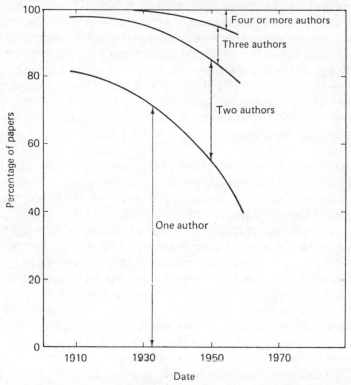

Fig. 4. Incidence of multiple authorship as a function of date. The data, which are from *Chemical Abstracts*, show an accelerating tendency towards more authors per paper.

Source: Price, *Little Science, Big Science*, p. 88.

seems to be a growing class of low-scoring or even 'fractional' authors who are necessary to back the prolific elite who lead research groups and teams.

Citations

One possible way suggests itself of refining crude paper counts so as to allow for differences in quality between individual papers. Could one find how much any given paper is used by other workers after publication? An important paper is likely to have a higher usage than a trivial one.

In practice, it is not easy to find how much a paper is read. One can, however, make use of the scholarly habit of acknowledging sources by quoting references to find how often a given paper is cited by other authors in their papers. This has become practicable with the appearance of the *Science Citation Index*, which lists citing papers under cited papers. For instance, one could look up in the 1970 volume a given paper published by J. Bloggs in 1962. Under the entry for that paper are listed the papers published in 1970 in which J. Bloggs (1962) was quoted as one of the references. With the *Science Citation Index*, one can thus do what is in a way the converse of following a series of references backwards in time – one can follow citations forward in time.

Using the *Science Citation Index*, one could do citation counts for all the papers published by a given author, or, for that matter, by a research group or an institute. As an academic exercise, the results could be used to 'weight' the crude count of papers published. But a very imperfect weighting it would be, for a number of reasons. Citation does not necessarily fully and faithfully reflect usage. In writing papers, scientists might, whether intentionally or not, preferentially quote their own papers, those of their colleagues or those of particularly well-known names. Some papers are rapidly superseded not because they are unimportant but partly because of their very importance. Others have the luck not to be improved upon and therefore continue to be quoted.

Consider, as an example, the *Science Citation Index* for 1967. In the list of the fifty most cited authors[7] appear two of the Nobel Prize winners for 1969, the physicist Murray Gell-Mann, ranked at number 6 with 942 citations, and the chemist D. H. R. Barton at number 41, his papers having been quoted 632 times in articles published during 1967. This looks promising. Citation counts seem to be of some value as at least a preliminary guide to quality.

However, consider the top of the table. Easily at the head of the popularity poll is O. H. Lowry with 2,921 citations, well ahead of B. Chance in second place with 1,374. More than 2,000 of the Lowry citations are to a paper in the *Journal of*

Biological Chemistry in 1951, which describes a technique for protein measurement by means of a combination of two well-known colour reactions. The combination gives a more intense colour with a given amount of protein and provides a sensitive and convenient way to estimate the total amount of mixed protein present in a sample. Clearly it was still in very wide use in 1967. The citation index is rather hard, though, on the three co-authors whose names appear with Lowry's on the paper that hit the jackpot, since it lists only the first-named author of cited articles. Furthermore, nobody is likely to claim that the new technique represents major scientific breakthrough.

Not that 'mere technique' is always to be looked down on. Nobel Prizes have been given for inventing new techniques. A notable case is the chemistry award in 1952, which went to A. J. P. Martin and R. L. M. Synge for introducing partition chromatography.[8] Their original paper on this, 'A new form of chromatogram employing two liquid phases', published in the *Biochemical Journal* in 1941, gave rise to an enormous amount of work, using partition chromatography for separations and analyses as well as developing other forms of chromatography from it. But this is not reflected in recent citations of the original paper, which numbered only 17 in 1967. The great tidal wave of publications on chromatography has left it behind.

What does all this show about possible ways of measuring quality? Are they attempts to quantify the unquantifiable? The kind of anomalies which have just been illustrated drive one point home most emphatically. Numerology of the kind discussed in this chapter is interesting but should be used only with the utmost circumspection. In particular cases it could be wildly misleading. Is it in fact likely to be misused? Is there a danger that it might supplant better criteria for determining growth rates, allocating grants and deciding promotions? If so, should work on these lines be discouraged? Or will it do to adopt the ethical neutralist position (p. 19), doing the work just to get the information and leaving it to others to decide how, if at all, it is to be used?

3 Hypotheses in Science

Why consider methodology?

Quite apart from sheer academic curiosity, various possible reasons present themselves for trying to describe and understand the nature of scientific method. First, there is the hope of improving the performance of practising scientists in research by giving them methodological rules to follow. Knowing something about the theory of what they are doing might help them to do it more effectively. It is on this kind of supposition that schoolteachers, for instance, as part of their training are taught something about educational theory, and the value of this, though not beyond all question, is at least widely accepted. The same cannot, however, be said of science. Most science students emerge from their formal education quite unscathed by philosophical consideration of methodology. How much of a pity this may be is a matter on which opinions vary.

Second – a related point on a less exalted level – there is the hope of help in teaching science at schools, colleges and universities. Asked to state the objectives of science teaching, many teachers would include something about 'scientific method', or even, more grandiosely, 'The Scientific Method', suggesting something that is relatively simple, well defined and widely known. It is disconcerting, therefore, to find that much of what is said and written about it turns out to be not derived from any well-established tradition of scholarship but personal opinion, often naive and sometimes idiosyncratic.

This has implications outside science, for there is a third and even wider context in which the issue can crop up: the use of scientific method outside the natural sciences in dealing with social problems. The claim of social sciences to 'scientific' status was briefly discussed earlier (p. 24). However legitimate it may be, doubts were raised about how desirable it is to base practical policy on 'scientific' study of social matters. Before this matter can be taken further, it is necessary to be clearer about what it is that social sciences are supposed to have borrowed from the natural sciences. What methodological prescription do they copy in their attempts to emulate the success of the natural sciences and to achieve in the social sphere the same kind of certainty of knowledge of the phenomena and power of control over them that we have in the physical sphere?

We might as well start with the majority view. Ask people what they conceive 'scientific method' to be – what it means to tackle a problem 'scientifically' – and what sort of answers is one likely to get? A few may be more sophisticated but the majority are likely to raise points like the following.

1. Science is *empirical, based on fact*; a scientist faced with a problem starts with systematic observation and/or experiment to gather as many pertinent facts as possible.

2. Science is *logical and objective*; it is not mere opinion; it does not operate by subjective impressions or wayward fancies, nor is it swayed by personal sentiments, but it treads a rigorous path, so that every right-minded person, once he understands, must assent.

3. Because of the above – because science operates on facts with logic its conclusions are *certain*. When something has been 'scientifically proved' there is no more room for argument, for the authoritative answer has been reached.

4. Science gives answers which are not indefinite and woolly but definite and precise; frequently they are exact and *quantitative*. Whenever possible the scientist measures, so that he can avail himself of mathematical reasoning, the most powerful form of logic yet devised.

5. Science is often *analytical*; it explains things by reducing them to simpler components – the 'nothing but' type of explanation. Thus a symphony becomes 'nothing but' a complex series of waves in air, or a person 'nothing but' a complex assemblage of atoms.

6. Partly as a consequence of this type of approach, science is a highly *specialised* activity, the realm above all others where it is the voice of the expert that counts.

Put thus starkly, the picture may seem exaggerated or even caricatured, but each of its facets is widely held. Versions of it colour a great deal of what is said and written about science.

Anybody who keeps his eyes and ears open as he reads books, magazines or newspapers, listens to radio or watches television can soon collect plenty of examples of the kinds of points made above.

Does it matter much if the picture is misleading? Not much, if the only question at issue is how scientists operate inside their laboratories. How the backroom boys do their work in the back room is not of momentous concern to most people outside. But the issue is wider, because of the attempted or supposed transfer of method from the field of physical phenomena to social matters. If our society comes to be run by technocrats according to allegedly 'scientific' methods, there is good reason to look more closely at the methodological recipe – especially since many people are not enthusiastic about the technocratic way of running things.

Fears of technocracy
One unusually articulate expression of what it is that people dislike and fear about technocracy is given by T. Roszak in his book *The Making of a Counter Culture*. His 'counter culture' is the contemporary youth culture with its interests in the psychology of alienation, oriental mysticism, psychedelic drugs and experiments in communitarian living. This counter culture, he says, 'comes closer to being a radical critique of the technocracy than any of the traditional ideologies', and he devotes most of his book to counter culture writers such as Marcuse and Goodman.

What concerns us most here is how he identifies and characterises the 'enemy', the technocracy. A technocracy, he says, is dominated by the 'reductive rationality' which the objective mode of consciousness dictates. It is a 'society in which those who govern justify themselves by appeal to technical experts who, in turn, justify themselves by appeal to scientific forms of knowledge. And beyond the authority of science, there is no appeal'.[1]

Roszak does mention criticisms by writers such as M. Polanyi and T. S. Kuhn (see Chapter 4) of the idea that science is a purely rational and objective activity[2] but his treatment as a whole does in fact view it in that light, as his usage of the word makes quite clear. 'Scientific knowledge is not just feeling or speculation or subjective ruminating. It is a verifiable description of reality that exists independently of any purely personal considerations. It is true . . . real . . . dependable. It works. And that at last is how we define an expert: he is one who *really* knows what is what, because he cultivates an objective consciousness'.[3]

'As the spell of scientific or quasi-scientific thought has spread in our culture from the physical to the so-called behavioural sciences, and finally to scholarship in the arts and letters, the marked tendency has been to consign whatever is not fully and articulately available in the waking consciousness for empirical or mathematical manipulation to a purely negative catch-all category (in effect, the cultural garbage-can) called the "unconscious" . . . or the "irrational" . . . or the "mystical" . . . or the "purely subjective". To behave on the basis of such blurred states of consciousness is at best to be some species of amusing eccentric, at worst to be plain mad.'[4]

The fundamental problem, says Roszak, is not political but transpolitical. Workers' control, the ideal of the French General Strike of May 1968, 'would amount to nothing more than broadening the base on which the technocratic imperative rests'. 'Centralised bigness breeds the regime of expertise, whether the big system is based on privatised or socialised economics.'[5]

Is Roszak tilting at windmills? Not entirely. Some aspects of decision-making in modern societies have already moved far in the direction he describes. Consider, for instance, the debate over the choice of a site for the third London airport, which brought to the surface a good deal of anxiety over the encroachment of technocracy. It was obvious that whichever choice was made would be painful to many, and one of the most elaborate cost–benefit exercises ever mounted was undertaken to help with the decision. The attraction, in theory, was that for once the day would not be carried by the loudest, best run and best financed pressure group; instead, all the relevant factors would be identified and taken rationally into account so as to find what would be best in the interest of the community as a whole, weighing all the pluses and minuses on the common quantitative scale of cost. How else could one combine the many considerations bearing on so complex a situation? Not only the construction costs at each of the four short-listed sites had to be estimated but also the accessibility to passengers and freight; the loss of homes and of farming land; the number of households within the various contours of noise intensity, considering gains near existing airports as well as the loss near the new one; the increase in jobs and trade which for some people counterbalances the noise nuisance ('That's not noise – that's a second car for the family'); the accident risks to those on board aircraft on the one hand and to those on the ground on the other.

The main fear among the public was, of course, that at the end of it all, out would pop some set of numbers from a computer which would given the unique correct answer. This might avoid bitter argument and achieve the ideal of reaching a conclusion to which everyone could, or at least should, assent (p. 25). But it would also supersede the delicate business of reconciling different interests which is held to be the essence of the political process.[6,7] Unique solutions are the negation of politics. They remove choices from the sphere of human discussion and judgement. Decisions come to look like cold, impersonal and mechanical ones. Although it is not true that cost–benefit analysis ignores non-materialistic standards, there is suspicion that, all the lip-service notwithstanding, the less easily quantifiable factors will be given less than their due weight. Some factors remain quite intractable to quantitative assessment. There was, for instance, the question of preserving Stewkley and some other villages with old churches near the site at Cublington, Buckinghamshire, as against the threat to the Brent geese at Foulness on the Essex coast. One witness equated the destruction of wildlife with the destruction of Ely Cathedral, clearly valuing it more highly than Stewkley. But not everyone need agree with this valuation.

In the event, the worst fears of computerised decision-taking were not realised – partly in that the Roskill Report[8] set out the pros and cons in a balanced way; more clearly in that the Commission were not able to report unanimously, Professor Colin Buchanan dissenting against the majority choice of Cublington in favour of Foulness; and most clearly of all in that the Government did not accept the majority recommendation. Whether this represents a triumph for humanity and political common sense over technocracy, or a setback in progress towards better and fairer modes of decision-taking, remains an open question.

Method in natural science
Our main purpose here is to consider scientific method in its
original home in the natural sciences, so we must now leave
the question of the use – or misuse – in the social sphere of
what is popularly conceived to be a scientific approach to take
a closer look at how physical scientists and life scientists
operate. From this it will emerge that the simple view described
above is inadequate and needs to be qualified in important
ways.

As an exponent of a somewhat more sophisticated view I
will take Sir Peter Medawar. Medawar won a Nobel Prize
for medicine in 1960 in recognition of his work on immunology;
until 1971 he was Director of the National Institute for Medical
Research in London. Thus he speaks as a practitioner of science
of great distinction. 'He ought to know, if anybody does',
some people might say. How justified they are in saying it is
another matter. Clearly a Nobel Prize for science is not a
negligible qualification for pronouncing on scientific method.
But distinguished practitioners are not necessarily best placed
to analyse the nature of the activity they practise. Perhaps they
operate by instinct, intuition and 'feel' rather than by consciously formulated principles. Asking scientists about scientific
method might be like asking birds about aerodynamics. Their
very involvement in the activity may hinder them from standing
back to take a detached view of it. Poets and composers do not,
after all, invariably make the most perceptive critics of literature
and music. Anything they might care to say about writing or
composing is valuable evidence, certainly, but not unquestion-
able authority.

In any case, there is no unanimity about methodology among
scientists themselves. Many of them lose no sleep over such
abstract issues, and philosophical subtleties soon drive them to
exasperated impatience. It is certainly not the case that conscious
attention to explicitly formulated methodological principles is
necessary for success in practice: philosophical naivety is no
bar to winning a Nobel Prize.

Medawar himself, however, is anything but philosophically
naive. Much of his conceptual framework derives from Sir
Karl Popper,[9] Professor of Logic and Scientific Method at the
London School of Economics until 1969. Medawar simplifies
it too far to satisfy professional philosophers but he presents it

with stylishness and wit. He reaches the nub of his argument[10] by presenting two contrasted conceptions of science in a fine passage which is given in Appendix 3. The two conceptions may be tabulated as follows:

Imaginative, intuitive	Critical, analytical, sceptical, dispassionate
Resembles art	Antithetical to art
Need not be useful	Usefulness is a measure of success
Requires freedom	Stimulated by practical needs
Support should be for individuals	Support should be for team projects

Clearly it is the second view (right-hand column) which corresponds better to the popular view described earlier. The other view is less cut and dried, 'softer', more human, leaving more scope for creativity and also for fallibility. It allows for the conclusions of science not to be certain beyond all doubt and for scientists to operate otherwise than as logical machines, which is more in keeping with the wealth of evidence from the historical record and from observation of contemporary science.

Yet the 'harder' view cannot be entirely wrong. How are we to reconcile the two conceptions, so very different and apparently contradictory? Medawar proposes that they are not to be regarded as alternatives to be decided between but that both exist as different phases of the scientific process – 'two successive and complementary episodes of thought that occur in every advance of scientific understanding'. The common mistake has been, he suggests, to view science in terms of induction. Induction in logic means arguing from particular facts to general principles or laws, as distinct from deduction, in which the logical argument is from the more general to the more particular. Thus induction can be represented as follows

$$\text{facts} \xrightarrow{\text{logic}} \text{laws}$$

though the proviso must be added that the simplicity of the representation must not prejudge the issue of whether such a process ever takes place.

The conclusions reached by this kind of inductive procedure would be certain (or, to be more precise, as certain as the facts from which it starts). This would match up to that popular but mythical ideal of science as a body of indubitable knowledge based on established facts from which conclusions are drawn by impeccable logic. But science, according to Medawar, does not operate by induction. Rather, he suggests, it uses a two-stage procedure sometimes described as hypothetico-deductive. In the first stage a hypothesis is guessed at; in the second stage the hypothesis is checked by deducing consequences from it and comparing those consequences with observations. This scheme depends on a clear distinction between the origin and the testing of theories – the way in which theories arise on the one hand and on the other the critical scrutiny which causes them to be accepted or rejected. Only the latter depends on hard facts and rigorous logic; the former is imaginative and intuitive, a matter of guesses, hunches and inspiration.

Assuming for the moment that the hypothetico-deductive scheme is a fair representation, where does that leave the status of scientific knowledge as regards certainty? A few moments' thought might suggest that we cannot know what is certainly right but that we can know what is certainly wrong. If a theory leads to conclusions not in accord with the observations, it must be wrong; but if there is accord, the theory, though supported, is not conclusively proved because it remains possible that some alternative is the true, or at least a better, explanation. When science is pursued in this vein, attempts at falsification – attempts to *disprove* theories – become the main purpose of experiments. They enable us at least to rule out what is false and thereby approach truth, even if we can never know whether we have achieved it. 'I can now rejoice even in the falsification of a cherished theory, because even this is a scientific success', wrote one scientist.[11] The best established theories are those which have over a long period withstood a gruelling procedure of non-destructive testing.

Some people have indeed interpreted hypothetico-deductivism in the light of such 'naive falsificationism'. But in fact even falsification cannot be absolutely certain. Tests themselves involve assumptions about facts and theories; if a test gives a 'falsifying' result, it may be one of these that is wrong rather than the theory nominally under test.[12]

Thus scientific knowledge has a lower status, in terms of certainty, than mathematical knowledge. A set of mathematical propositions can be certain with all the force that logic can command. Euclidean geometry, assuming the truth of some general axioms about space, parallel lines and so on, can prove beyond all doubt that the angles of a plane triangle are together equal to two right angles and that the square on the hypotenuse of a right-angled triangle is equal to the sum of the squares on the other two sides. But this is *deductive* logic. All the particular conclusions were already immanent in the assumptions, and brainpower was needed only to draw them out in explicit form. As some mathematicians modestly say, mathematics would be unnecessary if we were not so stupid that we cannot immediately see the more distant implications of our assumptions. It seems that there is no comparable logic by which induction can operate. If one tries to increase the information content by going from particular facts to general principles, one cannot avoid introducing the possibility of error. One can only achieve the appearance of induction by the hypothetico-deductive stratagem, which weakens certainty into mere probability.

This conclusion may sound depressing for all except those who indulge in *Schadenfreude*, taking a malicious pleasure in seeing uppity science put in its place: what had looked like the firmest edifice built by the human intellect on a solid base is reduced to a shaky structure on shifting sands. That expression so common in everyday speech – 'it has been scientifically proved that . . .' – takes on a hollow ring. On the other hand, there are compensations from the point of view of the image, and in particular the self-image, of scientists. The comparison with fine arts can easily be made to sound over-pretentious, but it does enable them better to hold up their heads against cultural snobs who regard them as 'rude mechanicals' plying their humdrum trade cut off from the higher, finer things of life. It may even be that in some cases there has been a liberating effect on the creative potentialities of scientists, encouraging them to give freer rein to speculation than the imagined shackles of rigour would otherwise have allowed. Inductivism enjoins caution, whereas a hypothetico-deductive system promises big rewards for bold conjectures.

These abstract methodological considerations must now be related to some examples from the historical record to serve as illustrations and as touchstones of validity. Even those who regard themselves as ignorant of science can probably, if they let themselves dare, pick out of their fund of general knowledge enough about a few cases of scientific discovery to see how easily they can be fitted into the hypothetico-deductive mould. Here I will take as illustrations two outstanding accounts accessible to non-specialists: Arthur Koestler's *The Sleepwalkers* and James Watson's *The Double Helix*.

First example: the Copernican revolution
Koestler's theme is the revolution in astronomy in the sixteenth and early seventeenth centuries. It is easy to see from his account that he has been a journalist and novelist rather than an academic: it is wordy but eminently readable, with the characters vivid, perhaps larger than life. But the book is by no means negligible as a piece of historical scholarship. The title indicates the central argument: that the – almost literally earth-shaking – replacement of the old earth-centred picture of the universe by a sun-centred one did not come about solely by the operation of cold clear rationality.

Copernicus, contrary to the view promulgated in some boys' story-books of the Heroes of Science genre, did not spend long years assiduously amassing accurate measurements of the positions of the planets. His *Book of the Revolutions of the Heavenly Spheres* contains only 27 new observations. Nor did they surpass in accuracy what the ancients had achieved. His contribution came through trying to devise a better scheme to account for existing data. The picture of the universe accepted at his time derived from Ptolemy (second century AD) and consisted of a complicated system of circles with the earth at the centre. It accounted for the observed planetary positions to a high degree of accuracy, but there were discrepancies, some of them magnified by the passage of centuries. Copernicus developed an alternative system of circles with the sun at, or rather near, the centre. The accuracy of fit with the data was better in some ways but worse in others; there was not much to choose between the two systems overall on this count. The new one was somewhat simpler, in that it used fewer circles, but the difference was not enough to carry overwhelming conviction: 48 circles (by Koestler's count,[13] rather than the more commonly quoted 32) as against 80 in the Ptolemaic system. Copernicus' own lack of confidence probably helps to account for his delay in publishing. His book was not printed until 1543, and he did not hold a copy in his hands until he was 70, and on his death bed.

No question here, then, of starting with indisputable facts and inducing incontrovertible conclusions from them; rather, of trying out a new hypothesis or 'model' of the universe and using some rather complicated mathematics to see how well, with suitable tailoring of the details, it could be made to fit the data.

The same is seen even more clearly in Koestler's account of Kepler, which is the best part of his book (indeed, the rest of the book was written around it). By the first decade of the seventeenth century better data had been gathered by Tycho Brahe, who had devoted much effort to collecting more comprehensive and more accurate observations of planetary positions, using larger instruments capable of greater precision. Kepler, playing the role of theoretician to complement Tycho the master observer, performed prodigies of laborious mathematics to try out one possible orbit after another for the planet Mars. Each orbit represents a mathematical model or theory. Each had to be discarded in turn until eventually he found that an elliptical orbit would fit the data within the new limits of precision.

Galileo's observations with the telescope added support for the sun-centred picture, but the evidence was still not conclusive. The discovery of mountains on the moon showed that classical theory was wrong in supposing all heavenly bodies to be perfectly smooth spheres; but this information had no direct bearing on the problem of their motions. The satellites of Jupiter showed that there can be bodies revolving about centres other than the earth; but this was not tantamount to showing that the earth revolves about the sun. The phases of Venus, it is true, seemed decisive against the Ptolemaic system; but they were not decisive for the Copernican, since they were equally compatible with the compromise system in which moon and sun revolve round the earth while the planets revolve round the sun. Even with a telescope, one cannot see that the earth moves while the sun remains stationary.

In the revolution which gave birth to modern science, then, there was a great deal that was not a drawing of rigorous conclusions from well-established facts. The point is driven home by the paradox that it is by no means certain that our modern picture of the universe does not in some ways resemble Ptolemy's more closely than Copernicus', Kepler's or Galileo's. Direct comparison is not possible because the space we talk about nowadays is a different concept; we cannot say simply that on some given point Ptolemy was right and Copernicus was wrong. But we would surely not insist today that the sun is stationary at the centre of the universe. The universe observable by us extends for enormous distances in all directions from where we observe, so is it not justifiable to say that the universe we live in is more nearly an earth-centred than a sun-centred one? Admittedly, the dimensions of the universe are so immense that the earth–sun distance is quite insignificant by comparison; but merely to establish the point as an arguable one is significant.

So much for the view of science as the patient accumulation of bits of truth. Is it not rather the case that progress, if such it can be called, consists in successively replacing world-pictures by others which seem better but are also conjectural and fallible?

Second example: the double helical structure of DNA
Watson's book has little in common with Koestler's except for readability. It deals with a very different segment of science – molecular biology in the early 1950s. Short and pithy, it bubbles with the irreverent wit of a clever young man. Not a historical reconstruction but an eye-witness account by a participant observer, it relates how Watson and Francis Crick in Cambridge did the work which led to the double helical structure for DNA and earned them, together with M. Wilkins, a Nobel Prize in 1962.

What Watson and Crick did was essentially model-building. They built models not in any sophisticated sense of conceptual schemes or mathematical propositions but in the most literal, simple-minded sense – fitting together sticks, plates, balls and so on to make molecular models. Each model represents a hypothesis about molecular structure. There was much playing about with them, sometimes fiddling with details, sometimes trying a quite different line of approach. Perhaps the most crucial stage came when Watson, too impatient to wait for the metal models of the four kinds of base which DNA contains, cut some for himself out of stiff cardboard and began to shift them around on his desk top. 'Suddenly I became aware that an adenine-thymine pair held together by two hydrogen bonds was identical in shape to a guanine-cytosine pair held together by at least two hydrogen bonds. All the hydrogen bonds seemed to form naturally; no fudging was required to make the two types of base pairs identical in shape.'[14] This base pairing, it has now been firmly established, provides a replication mechanism vital for the biological role of DNA as a carrier of genetic information.

Considering this little piece of history in terms of the hypothetico-deductive scheme, there are two points to which attention should be drawn. First, the entire procedure is not necessarily carried out by a single scientist or group. Watson and Crick did the crucial model-building but they could not, of course, have done it in a data vacuum. An immense amount of experimental work by others, especially in chemical analysis and X-ray crystallography, provided the clues which guided them and eventually authenticated the solution they proposed. Such division of labour is not exclusively modern; the way Kepler used the data gathered by Tycho is comparable.

Second, if scientific theories do not arise logically from the observed facts, how do they originate? Hypothetico-deductivism leaves this question shrouded in all the mystery associated with expressions like inspired guesses or intuitive hunches. Extraneous factors, however unlikely or trivial they may appear, can on occasion help. Popper argues[15] that even myths and superstititions should not be derided because, ill-founded though they may be, they can act as sources of theories.

In Watson's case, there is a source which may not be as facetious as it appears at first sight. Through his book runs much mild humour about bacterial sex, foreign *au pair* girls and so on. Is this just light-hearted banter to set off the serious scientific theme, or is it perhaps a pointer that the serious theme itself is – as in so many other good books – the multifarious manifestations of sex? How could Watson, a young, inexperienced biologist, deeply interested in genes but frankly ignorant of and indifferent to physics and chemistry, contribute to the desperately complex problem of the structure of the DNA molecule? Partly because of an intuition about sex, of which base pairing is the expression at the molecular level. At one stage there was a serious choice to be made between two-chain and three-chain models. Watson describes how he tried to decide this on a cold evening train journey from London to Cambridge. 'By the time I had cycled back to college and climbed over the back gate, I had decided to build two-chain models. Francis would have to agree. Even though he was a physicist, he knew that important biological objects come in pairs.'[16]

4 Paradigms in Science

Normal science
Until the late 1960s, views like those of Popper and Medawar were probably the most popular among that minority of scientists who have any use at all for philosophy. Since then, however, a strong challenge has come from the views of the American scholar T. S. Kuhn. Kuhn entered the field via the history of science – he has a fine book on *The Copernican Revolution*[1] to his credit; his most influential book, *The Structure of Scientific Revolutions*, slipped into the academic world unobtrusively in 1962 but is now widely read and discussed by people interested in the philosophical bases not just of natural science but also of other branches of knowledge. An illuminating confrontation between Popperian and Kuhnian viewpoints can be found in a book edited by Lakatos and Musgrave;[2] an extract from Kuhn's own introductory contribution to this book is given in Appendix 4.

At first sight, there is much in common between Popper and Kuhn. Neither, for instance, sees scientists as gathering facts uninfluenced by expectations derived from a conceptual framework of some kind. Nor does either of them believe that there are rules for inducing correct theories from facts (p. 52). They take different views, however, of the ways in which scientific knowledge grows. According to Popper, knowledge grows through criticism; good research consists of making bold conjectures and then ruthlessly criticising them. Kuhn believes that this sort of activity occurs relatively rarely – only in those periods of scientific development which are called revolutions. In 'normal science', according to Kuhn, theories are not so much subjected to tests as used in order to solve puzzles; science, he says, usually operates within the framework of existing theory, which provides it with a powerful puzzle-solving tradition, an effective set of tools and techniques for doing research. Whereas Popper's science is always – or at least should always be – trying to overthrow tradition, Kuhn's is for most of the time exploiting its potentialities.

Presented with Kuhn's view, Popper admits that 'normal science' exists, but he deplores its existence.[3] He dismisses it as just bad science, done by workers who are not critical enough, perhaps because they have been badly taught. Before we decide whether we can afford to write it off as easily as that, we must look more closely at it to see whether it is not in fact, as Kuhn claims, an essential feature of the development of science.

'Normal science' is defined by Kuhn as 'research firmly based upon one or more past scientific achievements, achievements that some particular scientific community acknowledges for a time as supplying the foundation for its further practice.'[4] Such achievements are called 'paradigms'. As examples of actual scientific practice they are more than just theories; they include the body of accepted theory, together with its successful applications and the appropriate instrumentation. They form research traditions such as the Ptolemaic or Copernican in astronomy, the Aristotelian or Newtonian in dynamics, the corpuscular or wave theories in optics. The study of paradigms is what mainly prepares a student for membership of a scientific community. 'Because he there joins men who learnt the bases of their field from the same concrete models, his subsequent practice will seldom evoke overt disagreement over fundamentals. Men whose research is based on shared paradigms are committed to the same rules and standards for scientific practice. That commitment and the apparent consensus it produces are prerequisites for normal science, i.e. for the genesis and continuation of a particular research tradition.'[5]

The paradigm concept

Kuhn's notion of paradigms has passed into common usage. The trouble with it is that it lacks precise definition. In his 1962 book, Kuhn uses the word in a number of somewhat different senses; an eager student[6] has managed to distinguish no fewer than twenty-one slight variants. Kuhn himself admits to 'having lost control of the word'.[7]

The dictionary meaning is 'example' or 'model', but a Kuhnian paradigm implies more than just an individual achievement which gives a lead for further work. It is a research tradition, a line of thought which carries a set of assumptions and guides a group of scientists in the way they are to approach phenomena, in what terms to think about them and analyse them, what kinds of effects to look for, what types of experiments to do and what sort of methods to use. It provides a way of seeing problems and suggests what kinds of techniques are appropriate and what kinds of solutions are acceptable. Kuhn is regrettably unforthcoming with clear, specific examples from modern science and leaves a lot of interpretation to the reader. Put very simply, the sort of thing that is to be understood is as follows. Physicists study Newton's second law, $f = ma$, and learn to look at mechanical problems in terms of forces, masses and accelerations; they develop further forms of it for dealing with particular types of situations – for freely falling bodies, for pendulums, for coupled harmonic oscillators, and so on. In an analogous position for chemistry one might perhaps put the notion of elements combining in constant proportions by weight, interpreted in terms of atoms joining up to form molecules. In some kinds of biology, the cell concept could be seen as playing a similar role as a paradigm of a very basic, global kind.

A paradigm change constitutes a scientific revolution. Kuhn takes optics as an example. Today's textbooks say light is photons, entities which have some characteristics of waves and some of particles. In the nineteenth century, light was considered to be a wave motion, on the basis of experiments on interference, diffraction, etc. In the eighteenth century 'the paradigm for this field was provided by Newton's *Opticks*, which taught that light was material corpuscles. At that time physicists sought evidence, as the early wave theorists had not, of the pressure exerted by light particles impinging on solid bodies'.[8] The different paradigm suggested a different kind of experiment, a different sort of effect to look for.

Before Newton, however, there was no single generally accepted view, but a number of competing schools. 'Being able to take no common body of belief for granted, each writer on physical optics felt forced to build his field anew from its foundations. In doing so, his choice of supporting observation and experiment was relatively free, for there was no standard set of methods or of phenomena that every optical writer felt forced to employ and explain. Under these circumstances, the dialogue of the resulting books was often directed as much to the members of other schools as it was to nature.'[9]

This apparently unkind cut by Kuhn need not be taken too hard. Arguing about the fundamental structure of a field – about what are to be taken as its basic assumptions and proper rules of procedure – is not necessarily a bad thing. It does not rule out the possibility of making significant discoveries and inventions. But it does indicate a different stage of development of a branch of inquiry from that of a mature science. In a mature science, because the ground rules are accepted, scientists can get on with the job. Chemists rarely spend much time arguing out the evidence for the existence of atoms. If they did, they would make less progress in determining structures, synthesising new compounds, and so on. Being able to take atomic theory and its developments and refinements for granted, they have a powerful research tradition for tackling the problems raised by the existence of different kinds of matter.

One might wonder whether large areas of the social and behavioural sciences, by contrast, do not in some ways resemble more closely the state of optics before Newton. They are less powerful in prediction and in problem-solving; and they are also more prone to disputes between schools because there is more questioning about the proper modes of approach and about how to formulate the questions. Perhaps they can with some justification be described as being in a 'pre-paradigm state'.

Thus the acquisition of a paradigm can be seen as a sign of maturity in a field. It makes possible more esoteric research. It also has an effect on the way in which information is communicated within the field. The predominant vehicle becomes the brief article or paper rather than the book. When scientists can take a paradigm for granted, first principles can be left to textbook writers and research communications become articles addressed only to professional colleagues. A moment's thought shows that this is an inescapable necessity. The chemical literature, for instance, is quite voluminous enough already, without each paper having to go back to square one and explain all about atoms and how they combine. The other side of that coin, though, is that an inward-looking exclusiveness tends to develop. Because so much is taken for granted in the readers, the literature is inaccessible to the uninitiated. It is fashionable to grumble about the high degree of specialisation and the heavy jargon of scientists. How far, one wonders, is this culpable negligence, in that they are not trying hard enough to make themselves more widely understood, and how far is it a necessary consequence of operating within a powerful, highly developed research tradition?

For scientists, says Kuhn, authorship of books ceases to be a principal sign of professional achievement. 'The scientist who writes [a book] is more likely to find his professional reputation impaired than enhanced'.[10] The difference between this state of affairs and that which prevails in arts subjects is taken up elsewhere (p. 82).

Puzzle-solving

Kuhn compares normal science, proceeding under the guidance of a paradigm or a set of paradigms, with puzzle-solving. The significance of the comparison with puzzles is this: solving a puzzle does not necessarily have much value for anybody else, but the existence of a solution is assured in advance and the kind of solution that is wanted is known; for instance, a crossword sets the task of finding words of a given length with given letters in common. To reach that solution poses a challenge to skill, resourcefulness and ingenuity. 'One of the reasons why normal science seem to progress so rapidly is that its practitioners concentrate on problems that only their own lack of ingenuity should keep them from solving.'[11] Kuhn suggests that normal science is rather like this. Thus a chemist given a new compound knows that it must have a structure which, in most cases, he can find if only he is a good enough chemist.

While a paradigm or set of paradigms may be very successful in guiding puzzle-solving, it may also orient scientists away from socially important problems that are not currently reducible to puzzle form. Scientists tend to tackle problems they can solve rather than problems that 'need' solving from society's point of view: to investigate fundamental particles or build up complex heterocyclic molecules rather than find a cure for cancer. 'The insulation of the scientific community from society permits the individual scientist to concentrate his attention upon problems that he has good reason to believe he will be able to solve. Unlike the engineer, and many doctors, and most theologians, the scientist need not choose problems because they urgently need solution and without regard for the tools available to solve them.'[12]

Thus Kuhn suggests what is virtually a mechanism to account for the tendency to internalism in science. There is an impetus arising from within science to drive and guide advance of the kind in which each achievement builds on earlier ones in the field, and this is distinct from external influences which would orient science to tackle problems in the rest of society. If this impetus is allowed free play, science is less likely to be directly useful to society, but at least it is likely to be successful by its own standards. Medawar puts the idea succinctly in the phrase which forms the title of one of his books: science is 'the art of the soluble'. The context in which the phrase arises is his discussion of the difficulty laymen find in understanding why scientists seem so often to shirk the really challenging problems. 'No scientist,' he says, 'is admired for failing in the attempt to solve problems that lie beyond his competence. The most he can hope for is the kindly contempt earned by the Utopian politician. If politics is the art of the possible, research is surely the art of the soluble. Both are immensely practical-minded affairs.'[13]

This makes the best argument for adopting a *laissez-faire* attitude in science policy (cf. p. 26). Scientists themselves, it can be contended, must be the ones who know best what *can* be done, even if it happens not to be foreseeably useful outside science; thus the choice of research problems, decisions about strategy and where the next major thrusts should be made, are best left to the scientific community, unhampered and undistorted by questions of applicability for military, commercial, health, welfare or whatever practical purpose.

That much the academic-freedom-for-science lobby are justified in arguing, but in considering their case we do well to remember how much diluted it often is by high-sounding humbug. Medawar himself writes scathingly of 'that reasoning which champions Pure Research because, while it enables the human spirit to breathe freely in the thin and serene atmosphere of the intellectual highlands, it is also a splendid long-term investment. Invest in applied science for quick returns (the spiritual message runs), but in pure science for capital appreciation. And so we make a special virtue of encouraging pure research in, say, cancer institutes or institutes devoted to the study of rheumatism or the allergies – always in the hope, of course, that the various lines of research, like the lines of perspective, will converge somewhere upon a point. But there is nothing virtuous about it! We encourage pure research in these situations because we know no other way to go about it. If we knew of a direct pathway leading to the solution of the clinical problem of rheumatoid arthritis, can anyone seriously believe that we should not take it?'[14]

Scientific revolutions

Kuhn's picture of 'normal science' might seem relatively flat, unexciting and unglamorous, but by contrast the significance he attributes to revolutions goes very deep indeed. As is supposed to be the case in political revolutions, they change not merely things within the system but the system itself. Let us first see how they arise.

There is a paradox regarding the transition from normal to revolutionary science. Normal science resists novelty and tends at first to suppress it, but it is effective in bringing it about because 'anomaly appears only against the background provided by the paradigm'.[15] One cannot notice that something is not quite as expected unless one has detailed expectations. The quite small discrepancies between the planetary positions observed and those predicted by the Ptolemaic system, which gave rise to dissatisfaction in the later Middle Ages, are themselves a tribute to the accuracy of the Ptolemaic system. An existing system is not just a barrier to achieving a new one, inhibiting fresh thought by its grip on men's minds; rather, its existence, exploitation and working out in detail are prerequisites for change.

Resistance to novelty, therefore, is not just narrow conservatism stemming from mental rigidity. It would be foolish to treat every discordant observation as a falsifying counterinstance. Argon and potassium have atomic weights the wrong way round to fit into the periodic system of elements, but this contradiction did not cause the system to be abandoned.[16] In such situations, judgement must be exercised by scientists on the virtues of a given system. An apparent anomaly should not necessarily be allowed to overthrow the whole edifice; it is right merely to note it, with the provisional hope that the observations will in the future be amended, or explained, or explained away, or in some way fitted in.

This means that the simple form of hypothetico-deductivism is wrong in assuming direct falsification of a hypothesis in isolation by comparison with nature. Facts that do not fit do not by themselves cause renunciation of a paradigm. 'The decision to reject one paradigm is always simultaneously the decision to accept another, and the judgement leading to that

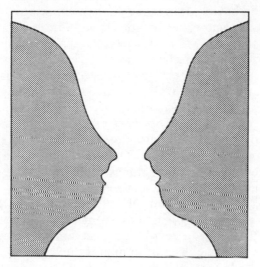

Fig. 5. A wine glass, or two faces in profile?

decision involves the comparison of both paradigms with nature *and* with each other.'[17]

But when a paradigm is renounced, the mental change is very radical. After a paradigm shift, one sees things quite differently – one could almost be said to live in a different world. A useful elementary prototype of what happens in a paradigm shift is a *gestalt* switch. Gestalt, a German word meaning something like 'form', is a concept used by psychologists to refer to perception of configurations as a whole rather than in terms of their discrete parts, and an example of a gestalt switch is what happens when, while looking at one of those 'trick' drawings that can be interpreted in more than one way, one suddenly 'sees' something quite different from what seemed to be there before.[18] The marks on paper that were first seen as a duck are now seen as a rabbit, or vice versa; or one can see either two faces in profile or a wine glass (see Fig. 5). The data do not change even in any minute particular, but the human mind can impose on them more than one pattern or meaning.

The Copernican revolution can be compared to this. Where Ptolemy had seen a system of circles centred on the earth, Copernicus looked at virtually the same data and saw a system or circles centred on (or rather near) the sun. The mental transposition having been made – a new thinking cap donned – new kinds of facts came to appear significant. Was it purely accidental that only after Copernicus did Western astronomers notice certain signs of change in those regions of the heavens which had previously been held to be immutable, far removed from the earthly centre of the universe and accordingly immune from the change and decay of earthly things? Technically, they did not have to wait for Copernicus to notice sunspots and new stars, or to find that comets move through regions of space more distant than the moon. 'The very ease and rapidity with which astronomers saw new things when looking at old objects with old instruments may make us wish to say that, after Copernicus, astronomers lived in a different world. In any case, their research responded as though that were the case.'[19]

But the Copernican revolution was something *very* special and unusual in the course of human thought. How many other events in the history of science fall into the same class? The biggest difficulty in applying Kuhn's thinking to actual cases is to decide how deep a change has to go to constitute a paradigm shift. What about the double helix discovery (p. 57)? Undoubtedly it was a very great scientific breakthrough, and epoch-making in the sense that it has led to a great deal of fascinating work on the molecular basis of genetics, the way in which genetic information is carried in 'coded' form in the sequence of bases along the DNA strands, and so on. But it can also be argued that getting at the double helix structure was just another problem in the architecture of molecules, an area in which a successful tradition stretches back more than a century. Is it perhaps better described as a piece of puzzle-solving – a particularly ingenious solution to a particularly important puzzle, no doubt, but still a step in the advance of the sort of successful research tradition that constitutes normal science?

The reader is invited to try to apply this kind of analysis to any episodes of science that he knows about. He will probably find that the notion of two qualitatively distinct kinds of activity, normal and revolutionary, is hard to maintain. It is difficult not to be convinced by direct inspection of science that contributions form a continuous spectrum, with really major changes very rare and minor ones the most frequent, but with intermediate ones occurring with intermediate frequency. Medawar says that 'the belief that great discoveries and little everyday discoveries have quite different methodological origins betrays the amateur'.[20] What Kuhn's analysis has revealed is perhaps best thought of as two important elements which are not sharply separated types of activity but represent complementary aspects contributing in varying proportions to given events. Could the double helix work be most satisfactorily described as two-thirds Kuhnian normal to one-third Popperian revolutionary science?

Education in science
One aspect of the analogy to puzzle-solving has significance in the educational context. The rules which are embodied in paradigms and which guide 'normal' research need not be explicit. Imagine trying to explain what a crossword is to someone who has never seen one. To formulate all the general principles would be immensely laborious compared to showing an example and working through it. The same is true in science. If a student ever discovers the meaning of terms like force and mass, he does so less from definitions than 'by observing and participating in the application of these concepts to problem-solution'[21]

Thus a fairly clear view emerges of the function of education in science: to initiate the student into a tradition by putting paradigms at his disposal and teaching him how to use them to good effect. By this means he acquires powerful tools with which to tackle problems.

This makes me wonder about some current orthodoxies in science teaching, especially at school level. The role of practical work and 'discovery methods' sometimes seem to me to be exaggerated.[22] It is suggested that pupils must not learn about magnetism or about floral mechanisms from a textbook or a teacher, but must play with magnets or dissect flowers to find out for themselves. The printed word becomes a bogey, sullying the open purity of mind with which direct observations should be made. One might think, listening to some enthusiasts for modern trends, that science was doing better before Gutenberg interposed his pernicious invention between the scientist and nature.

My belief is that this is over-reaction against the bad old days of thought-deadening rote learning. Of course direct contact with phenomena is essential, but the nature of that contact can be misconceived. Intelligent children at secondary school are passing beyond the stage where sheer exposure to empirical observations can pass for science education. They need to come to grips with the inevitable entanglement of observation with theory.

In the light of the above discussion of method, one can see how the misconception might arise. It could derive from the simplistic conception of science as inductivist, as the operation of an open mind on bare empirical facts. Can a mistaken conception of research method be a satisfactory foundation for education? To enable the young to benefit from the hard-won experience of the past, must one not introduce them to paradigms and successful puzzle-solving traditions?

Even if one adopts a more Popperian standpoint and concentrates on inculcating a critical attitude, one should consider carefully what is the best way to put pupils or students in a position to be critical. The few experiments they can do themselves, largely pre-selected and inevitably 'rigged', are a feeble basis for confident criticism. Would not the storehouse of observations and speculations recorded in books and articles make at least as powerful a stick with which to beat orthodoxy in the shape of teacher or textbook? No less than arts students, science students should learn to use literature critically. To give them the impression that books only present dogma misses one of education's most important lessons – that paper refuses nothing and that one should not believe all that appears in black and white on the printed page.

Rationality and progress
The ultimate conclusion of Kuhn's analysis seems profoundly pessimistic. It questions the possibility of progress towards truth.

Scientific revolutions of the past, says Kuhn, tend to be disguised by textbooks and popularisations of science. By writing history backwards, by referring only to 'that part of the work of past scientists that can easily be viewed as contributions to the statement and solution of the texts' paradigm problems',[23] they make science seem cumulative, a piecemeal process by which items are added to an ever-growing stockpile of knowledge and technique.[24] But there are no standards by which to judge whether paradigms develop cumulatively.[25] In the competition between paradigms for the allegiance of a scientific community, other things count besides problem-solving ability. There is also aesthetic appeal, whether the new view is neater or simpler.[26] Above all, there is the promise it holds for guiding future research. 'A decision of that kind can only be made on faith.'[27] 'I would argue . . . that in these matters neither proof nor error is at issue. The transfer of allegiance from paradigm to paradigm is a conversion experience.'[28] It is not necessarily the case that changes of paradigm carry science closer to the truth.[29]

To some, it seems that Kuhn is making science irrational. With the emphasis on conversion and commitment, research comes to appear as a matter of social psychology more than of logic. 'In Kuhn's view scientific revolution is irrational, a matter for mob psychology.'[30] Kuhn replies[31] that theory choice is not merely intuitive and that there are good reasons on which to base it: accuracy, scope, simplicity and fruitfulness. It can hardly be denied that there is progress in the sense of improved puzzle-solving. But the idea of closer and closer approximations to truth, to statements about 'what really exists and happens out there', may have to be recognised as incompatible with his view.[32]

Is all this too far from the traditional picture of science as an activity, eminently rational above all others, in which bits of truth are continually added to a growing accumulation, or at least in which absolute truth is approached ever more closely? Is it over-sophisticated, in the tradition of that kind of philosophy which proves incontrovertibly that what is obviously the case cannot possibly be so?

Time will tell how much of Kuhn's original thesis will find general acceptance. Perhaps it exaggerates the strength of commitment of scientists to existing theories and the degree to which they resist new patterns of thought. Provided the empirical facts seem clear cut, surprising discoveries are often accepted and assimilated by the scientific community without appreciable resistance, even if well-established theoretical positions have to be abandoned. The popular view of science as 'organised scepticism', appraising evidence in a detached and unemotional way, is not entirely a myth.

Certainly there is much about science which Kuhn's picture leaves unexplained, but there is no mistaking the deep impact it has made. It cannot be ignored by anyone who takes an interest in the nature of scientific activity. Its importance goes beyond philosophy into the sociology of science. Shared allegiances to paradigms have clear implications for the organisation of science as a social system and the cohesion of groups of workers in specialised fields of science. In the next chapter we turn to some sociological considerations.

5 What Makes Scientists Tick?

The information-recognition exchange model
One of the most important things to know about science is why people do it. What motivates scientists? What makes them tick?

Of course there is usually a mixture of reasons. Motivation 'is rarely pure, and never simple', as Oscar Wilde said of truth. Some of the more obvious possibilities soon spring to mind. There is sheer curiosity of the kind sometimes described as 'disinterested', in the curious usage of that word which denotes deep interest in how nature works. Some degree of altruistic concern for benefit to humanity usually colours this curiosity. Most scientists see science in part as a job, a respectable profession which provides an acceptable way to earn enough money to keep themselves and their families; from this point of view, it is not vastly different from other professions. Scientists may also see science as a way to gain esteem and recognition from their fellow men and women. In this, too, science is far from unique: actors, athletes and authors also seek recognition. One would, however, expect scientists to look for recognition more within their own profession, since the public at large is not in general well placed to appreciate their work.

It is this last factor – recognition by fellow scientists – which is stressed in W. O. Hagstrom's book *The Scientific Community*. Some of my comments here will be critical of this book's main thesis, but it should be said straight away that it is widely recognised as one of the most important books yet to appear on the sociology of science. Packed with evidence, in particular about the sorts of things that scientists say about their work, it clearly ranks as obligatory reading for anyone who takes a serious interest in the field.[1]

Hagstrom's central thesis is that science can be viewed as an exchange system in which gifts of information are exchanged for recognition from scientific colleagues. He puts forward, in short, an 'information-recognition exchange model' of science, if one wants to use the word 'model' in the more abstract sense, a conceptual scheme rather than a solid object. The information is presented as gifts in the sense that the results of research are normally published as papers in learned journals for which the authors do not expect payment. They get recognition in the first place by having their papers accepted for publication; a reputable journal does not publish everything that is submitted to it but seeks the advice of referees on whether the manuscript looks like a sound and worthwhile contribution. Further recognition comes through citation of their publications by other authors (cf. p. 44); through election to membership of societies such as the Royal Society; through the award of medals and prizes, up to and including Nobel Prizes; and through invitations to attend meetings, pay visits, give lectures and so on.

Before trying to assess the evidence for and against Hagstrom's theory, explanation and background should be given on three matters: first, as regards the different types of scientists that could be considered; second, about the general concept of exchange systems; and third, about the exact role that the exchange of information for recognition is supposed to play in the organisation of science as a social system.

Types of scientists
Large numbers of scientists are active today and there are clearly many ways in which they differ among themselves. What are the most significant ways in which one could classify them? Obviously, some scientists are better than others, so one can have good scientists and bad ones, or at least better and less good ones. One could also classify them by field, into such broad categories as physical, life and earth sciences, and also, if one wishes to include them (cf. p. 24), social and behavioural sciences. Again, one could classify them by sector of employment – university, industry, government, secondary school and so on.

For his own field work, Hagstrom used a sample of 92 scientists. Although he quotes the work of others relating to wider ranges of scientists, it is significant to note the restricted composition of the sample he himself chose to investigate.[2] As regards status, 76 were university teaching staff, 3 were other professionals (Ph.D.s engaged in research), 11 were graduate students and 2 were technicians; of the 5 universities represented, 4 ranked among the top 25 in the United States. As regards field, 75 were in the physical sciences, including mathematics, 13 in molecular biology and 4 in other areas. It is clear that Hagstrom's sample is composed overwhelmingly of high-grade, academic physical scientists, especially when one remembers that molecular biology is the kind of life science that is closest to physical science. Thus the sample represents only a small minority of all scientists (cf. p. 86). Of course one could assume that the high-grade, academic physical scientist is the arche-typical scientist, the ideal to which all other kinds of scientist aspire and of which they are mere imperfect copies. The feeling that this is so is in fact by no means uncommon. But, if any differences are noted, it would be appropriate to ask who is out of step with whom. Can it be that the many are out of step with the few?

Exchange systems

Some sociologists see the concept of exchange systems as a fairly general kind of explanation for the operation of social systems. Thus Storer[3] identifies four major social systems in society which can be regarded as exchange systems: the econo-mic, the political, the family and the religious systems. For the economic system the matter is obvious; money is exchanged for goods and services. For the political system the case is less obvious, but there are grounds for viewing the political process as one in which support is exchanged for influence; in other words, a leader gets power by letting others have a voice. Thus Crick says that politics 'can be simply defined as the activity by which differing interests within a given unit of rule are conciliated by giving them a share in power in proportion to their importance to the welfare and the survival of the whole community'.[4] With the family and religious systems, the ex-change model is less convincing, but Storer argues that marriage can be regarded as an exchange by the woman of sexual access for material support and protection, and that religion can be viewed as an exchange of prayer for salvation.

In formal analogy to these, Storer sees science as a fifth exchange system. In this case, information is exchanged for recognition or, according to the variation suggested by Storer, for 'competent response' from professional colleagues.

There is no need to enlarge here on the weaknesses of these exchange models, which are obvious enough, especially in the cases of marriage and religion, where the terms of the supposed exchanges are likely to be greeted by many with smiles, if not contempt. But there is one intriguing feature of exchange system theory which is worth noting. For the systems to work properly, they have to be kept separate and autonomous. Society disapproves of mixing them. Mixing the economic and political systems would mean using money to buy political power, or political power to get money, both of which are denounced as corruption. Mixing the economic and religious systems might mean trying to buy salvation with 'conscience money', an idea which is likely to raise titters. Mixing sex with the power or money systems would be called respectively rape or prostitution – both words which are emotively loaded in an unfavourable sense.

In the case of science, preserving the autonomy of the exchange system means keeping it free of non-scientific influences such as political and economic pressures. This, it is easy to see, is a re-formulation of the terms of the debate about internal and external factors and the extent to which the latter actually influence scientific development or should do so (pp. 27). Phrases about the 'prostitution' of science for commercial or military ends, which rise so easily to internalists' lips, now take on a fresh meaning: the analogy to the more literal meaning is really quite a close one.

A reinforcement role

The role which Hagstrom ascribes to the information-recognition exchange model is one of social *control*. It is in part because scientists desire recognition, he argues, that they adopt the goals and norms of the scientific community. The overriding goal is the extension of knowledge, and the norms are the guidelines which regulate behaviour in ways that are supposed to be conducive to the attainment of that goal – to assess evidence independently of the nationality, colour, creed or status of its author, to refrain from making fraudulent or exaggerated claims about one's own work, and so on.

Social control is to be distinguished here from 'socialisation', a word which in this context has nothing to do with left-wing ideologies but refers to the fact that scientists as individuals acquire and 'internalise' – make their own – the dominant values, goals and norms of the scientific community. Scientists become highly socialised through education. Science students are self-selected to begin with; young people who are out of sympathy with the aims of science are unlikely to opt for science. During education, their exposure to examples of scientific practice and their contacts with teachers and with fellow science students develop their commitment. Efficient socialisation of most recruits to the profession is necessary for its smooth functioning. If it were *fully* effective, no additional control would be needed. Each scientist would conform to scientific values and norms just because that is what he as an individual wanted to do. Hagstrom, however, criticises the view that no superimposed control is necessary, calling it 'naive individualism'. He argues that the information-recognition system reinforces and complements socialisation by a system of rewards and punishments, the giving or withholding of recognition by scientific colleagues.

Evidence: what scientists themselves say
With that much in the way of explanation and background, we can now consider how far the available evidence backs the theory of social control in science by an information-recognition exchange system.

What kinds of evidence could one bring to bear? Most obviously, one could ask scientists themselves. And here the theory meets its first difficulty, for it is clear that many if not most scientists, questioned directly, would deny being motivated primarily by the desire for recognition. A questionnaire would probably produce strong evidence *against* the theory. Scientists would be more likely to stress the genuineness of their desire to advance knowledge and doubtless many would be sincerely distressed at the suggestion that what they are really after is something as selfish as recognition for themselves personally.

It is true, of course, that the motives people express openly are not always the operative ones. People commonly hide the real motives – from themselves almost as much as from others. Still, it can give no comfort to supporters of the theory that scientists themselves do not readily assent to it. Hagstrom himself is forced to say that 'the very denial by scientists of the importance of recognition as an incentive can be seen to involve commitment to higher norms, including an orientation to a scientific community extending beyond any particular collection of contemporaries' (Appendix 5). Does this explain the facts, or does it merely explain them away?

Evidence: priority
The most powerful evidence in favour of Hagstrom's thesis comes from the high value scientists set on establishing priority for their discoveries. Medawar puts the point in characteristically spirited fashion and sees here a difference between creativity in science and in art. 'In science, what X misses today Y will surely hit upon tomorrow (or maybe the day after tomorrow). Much of a scientist's pride and sense of accomplishment turns therefore upon being the *first* to do something – upon being the man who did actually speed up or redirect the flow of thought and growth of understanding. There is no spiritual copyright in scientific discoveries, unless they should happen to be quite mistaken. Only in making a blunder does a scientist do something which, conceivably, no one else might ever do again. Artists are not troubled by matters of priority. ... Wagner would certainly not have spent twenty years on *The Ring* if he had thought it at all possible for someone else to nip in ahead of him with *Götterdämmerung*.'[5]

Sociologists of science have for some time been struck by scientists' intense concern over priority, by the devices they use to claim it and by the bitterness which disputes over it can engender. History is full of illustrative examples. For instance, Galileo first announced some of the discoveries he made with his telescope in the form of tortuous anagrams, so as to safeguard his priority without as yet disclosing what it was that he had observed.[6] Today there are various devices for staking a claim quickly to research results which, in many cases, are not yet ready for full publication – for instance, the publication of preliminary notes in journals or of abstracts of papers to be presented at forthcoming meetings of learned societies. Nor are priority disputes a thing of the past; on the contrary, they are common occurrences in the lives of research workers. Hagstrom reports[7] that at least 9 out of 79 informants admitted to having been involved in questions of disputed priority, either as culprit or as victim.

The belief that scientists are driven purely by an unselfish desire to advance knowledge appears, then, to be too idealistic and over-simplified. A particularly striking instance is reported by Watson with his usual candour. When he and Francis Crick had satisfied themselves that the latest DNA model put forward by Linus Pauling in California must be wrong, they 'went over to the Eagle. The moment its doors opened for the evening we were there to drink a toast to the Pauling failure. Instead of sherry, I let Francis buy me a whisky. Though the odds still appeared against us, Linus had not yet won his Nobel.'[8]

The strong evidence for the importance attached to priority is revealing, but it should not blind us to the fact that science is not unique in this respect. Mountaineering and athletics, for instance, are comparable: the second conquest of a peak carries less glamour than the first, and each athletic record can be broken only once.

One alternative: the pleasure of research
When one tries to think of alternatives to the information-recognition exchange system, one possibility that springs to mind is that scientists just like doing research. 'Mucking about in labs' gives them pleasure in the sort of way that 'mucking about in boats' gives pleasure to some people, and the sheer solving of puzzles is a satisfaction of its own. Scientists doing research, in short, are doing what they like doing anyway. Their behaviour is accounted for by self-selection and socialisation during upbringing and education.

To the extent that there is truth in this view – and no doubt there is some – the control system may have a real function to play in driving the scientist to communicate his results to other scientists. Without communication there can be no scientific community; a scientist in one laboratory cannot build on the results of work elsewhere unless he gets full and reliable information about that work. But many researchers feel that getting results ready for publication is less gratifying an activity than actually doing research. Editors of journals tend to be a fussy lot. They like to have the loose ends reasonably well tied up, which might involve laborious and relatively dull experimental work in clearing up details. Manuscripts have to be written concisely to conserve journal space – and readers' time! – and should follow editorial conventions which vary from journal to journal and sometimes seem perversely arbitrary. It is in the interests of science as a whole, therefore, that there should be an incentive such as the desire for recognition to encourage scientists to tackle the annoying trivialities involved in publishing.

One might think that there is an inconsistency in the argument here. If discovery is fun and writing up results is a chore, can the recognition which follows publication be an important motive? Is this not like arguing that industrialists are driven by the profit motive but admitting that they prefer the sheer fun of making goods to the chore of selling them and collecting the money?

The apparent inconsistency disappears when one remembers what precise function the information-recognition system is supposed to play. It is put forward as a system of social control superimposed on socialisation. Thus the intrinsic satisfaction of doing research is not really an alternative to the desire for recognition but additional to it.

Another alternative: extrinsic rewards

Another and perhaps even more obvious alternative to consider is that scientists want the same kinds of rewards as most people who follow careers, namely position and money. These are 'extrinsic rewards' – extrinsic, that is, to science, the rewards appropriate to the outside world rather than the scientific community. Hagstrom calls this the 'contractual theory', because it supposes scientists to behave as though they were on contract to those who pay their salaries.

How different is this really from saying that scientists seek recognition? Academic appointments and promotions depend largely on published research output. Recognition and advancement therefore go hand in hand, and it is difficult to disentangle the two and decide which leads to the other. Hagstrom's case depends on recognition coming first and good appointments following from it, rather than the other way round.

This is certainly the way things *should* work. Acceptance of a paper for publication is supposed to depend on the content of the paper, not on the status of the author. In practice there are doubtless deviations from ideality. An often-repeated story concerns a paper by Lord Rayleigh at a time when his reputation was established; it was rejected when submitted with his name inadvertently omitted, but accepted when the identity of the distinguished author became known.[9] Of course, the mere fact that the story is considered worth telling shows that the incident is regarded as irregular. Nevertheless, in a number of more or less elusive ways, it is probable that the position held by a scientist does influence signs of recognition accorded him. As regards appointments in particular, a version of the 'Matthew effect' (p. 42) is almost bound to operate: the fact that a scientist already holds a good appointment is likely to stand him in good stead when he is considered for a higher one. Nothing succeeds like success.

Hagstrom gives every appearance of wanting to play down the importance of extrinsic rewards. If the behaviour of scientists were significantly influenced by considerations of money paid by non-scientists for non-scientific reasons, this would threaten the autonomy of science, mixing the exchange system appropriate to it, according to the information-recognition model, with factors that properly belong to other systems – 'prostituting' science, in short. Naturally Hagstrom is not blind to the fact that in the real world the purity of science is sullied in this way, especially outside universities. He quotes data[10] which show that scientists in industrial research establishments publish less than academic scientists and that for them papers are less important for promotion. This might easily have been guessed, and indeed it is quite right and proper. Industry has other aims than to contribute to the international pool of knowledge. Even those who heartily dislike the profit motive must admit that industrial research laboratories are not the appropriate setting for work with no apparent relevance to the operational objectives of the organisation. The same applies to nationalised undertakings where taunts about the profit motive do not apply. Taxpayers and voters would justifiably disapprove if electricity, railway and hospital undertakings concentrated their efforts on filling the pages of learned journals rather than improving electricity, rail and health services.

Even within universities, there are things that matter besides doing research and publishing the results. Most university scientists spend appreciable proportions of their time teaching students, even though this activity does not increase their prestige in research. Do they teach for money but do research for recognition? Many university scientists also do a good deal of administration and for some this becomes a major, if not *the* major role. They could then, from the point of view of research, be dismissed as 'unproductive'. Yet there are probably quite a few universities, and certainly many faculties of science and technology, which would soon grind to a halt without them.

The information-recognition exchange model hides the fact that there are two kinds of recognition which are very different and are often in competition with each other. They can be called the 'cosmopolitan' and the 'local'.[11] 'Cosmopolitan' refers to recognition by the community of scientific colleagues, which is likely to be scattered through all or most of the advanced nations of the world, numbering probably thousands if one thinks of the general field, a few dozen actively engaged on the particular topic. 'Local' refers to recognition which is geographically restricted and comes from other members of the organisation in which the scientist works, be it university, institute or company. Most of the local colleagues are not likely to belong to the same discipline or speciality. The recognition, the respect and the rewards they can offer are clearly different in kind from what fellow specialists can give, but they are equally clearly not negligible in importance. Yet the information-recognition exchange model takes account of the first kind only.

There is a further qualification to be made regarding the model, and that concerns the extent to which it is characteristic of science. Is it not a description of the system within which any academic operates? Not only scientists but also historians, classicists, archaeologists and so on strive to win the esteem of fellow scholars by the quality of their learned publications. To the extent that they are dedicated to scholarship they are content to accept such recognition in part-replacement of worldly influence and financial rewards. The information-recognition exchange model is perhaps better thought of as characteristic of the academic rather than of the scientific community. It gives what is within limits a reasonable description of academic scientists but not of the many scientists of other kinds. The label sticks to the academic scientist more because he is an academic than because he is a scientist.

Of course there are differences between science and arts academics, and one in particular is worth mentioning here. It concerns the relative roles of books and papers as vehicles of publication. As has been noted (p. 63), in science it is papers rather than books which are the normal means of announcing results to the world. This is less true in arts, where a man tends to be known by his books. When historians consider a senior appointment or a promotion, the first question they are likely to ask about any candidate is 'What has he got in hard covers?' In science, the corresponding question would be 'How many papers has he published?' (cf. p. 40); and a book on the list of publications is likely to elicit the suspicious question, 'Is it a genuine contribution to scholarship or only a textbook or popularisation?'

The academic disdain of textbooks and popularisations is consonant with the information-recognition exchange model, because it is with books like these that there is a chance of appreciable 'extrinsic rewards' in the form of royalties. A tome heavy with scholarship is unlikely to be a money-spinner. It could therefore be held that textbooks and popularisations are written for money[12] and 'prostitute' science just as much as research aimed to increase company profits. Yet, on the other hand, a wider dissemination of knowledge about science is generally held to be a most desirable aim. Given the obvious fact that science and technology are mainsprings of change in modern societies, a well-informed public is perhaps the best guarantee that their enormous potentialities will be directed in the most beneficent way. Surely, therefore, good popularisation should be regarded more as a duty of the scientific community than as an aberration from its true purpose. Does this not amount to yet another illustration that trying to keep science 'pure' by isolating its exchange system from others turns out in the event to be not noble and idealistic but just plain anti-social?

Teamwork

Another complication regarding the information-recognition model concerns teamwork. Hagstrom's analysis bases itself initially on the independent worker free to choose his own problems and the way to tackle them; but the proportion of scientists who work quite alone in this way is decreasing. How does the information-recognition scheme apply when scientists collaborate?

It is easy enough to extend the model to cover free collaboration of a scientist with one or a few of his colleagues. There is then joint authorship of any publications that ensue and the recognition is shared. The same happens when a university scientist supervises a research student, which is how much university research gets done. There is then the additional complication in planning the work that it has to be chopped as neatly as possible into thesis-sized chunks, but when it comes to publication, it is again usual for both supervisor's and student's names to appear as authors on the papers that report the results.

But what if research depends on some really big machine such as a radiotelescope or a particle accelerator? Then work cannot be done without involving a large number of people. Hagstrom's diagnosis is that scientists then split into two roles – administrators on the one hand, technicians on the other. 'Leaders necessarily become politicised, oriented to obtaining funds and access to facilities and co-ordinating the efforts of others. Technicians become means-oriented, interested in performing their specialised skills for extrinsic rewards and uninterested in the recognition given by the scientific community for the attainment of scientific goals. If this occurs, the information-recognition exchange theory of scientific organisation is no longer applicable. Control is exercised by hierarchical authority within research groups and by political powers outside them. Scientists become more interested in their particular organisations and in the reactions of politically powerful leaders than in the opinions of the wider scientific community. Consequently, the complex organisation of science leads to disorganisation – disorganisation in terms of the information-recognition exchange theory of organisation' (Appendix 5).

A good deal of room is left for debate about the adequacy of this diagnosis and the distaste with which Hagstrom seems to view the trend he describes. Consider, for instance, the technician role. The phrase 'technician mentality' has a derogatory ring, suggesting people who are deployable for purposes decided by others, prepared to put their skills at the disposal of anyone who will pay, like mercenaries who will fight for any regime, however repressive or corrupt. But there is a sense in which it might be a good thing if scientists did regard themselves as more deployable. The fields that are prestigious by scientific criteria do not necessarily coincide with those that are most useful socially. Synthesising exotic new molecules carries more academic *kudos* than tinkering with old ones to make materials with useful properties. Would it not be good if qualified people were more flexible and willing to let themselves be 'deployed' into fields picked by criteria external to science?

Then again, consider the curious sense in which 'disorganisation' is said to set in. It is 'disorganisation' only in terms of the information-recognition control system, not in the ordinary sense. Indeed, it sets in just when *organisation* in the common sense begins – when scientists become organised in groups and teams instead of acting as independent individuals. This kind of organisation is necessary whenever science becomes big; and that means not only in industry or mission-oriented agencies where teams are deployed to attain objectives specified by 'political powers outside', but also in the purest of pure research, such as high energy nuclear physics. Running big particle accelerators requires a high degree of planning and co-ordination. When there is an experimental run at CERN, the European Centre for Nuclear Research at Geneva, the scene is said to be like the NASA mission control centre at Houston during a moon-shot.[13] The difference between the two kinds of 'organisation', therefore, does not correspond to the distinction between basic and applied work but rather to that between little science and big science.

Moreover, it seems highly dubious whether Hagstrom is right in suggesting that the kind of organisation which 'disorganises' information-recognition control necessarily leads to low commitment. High energy nuclear physics is often attacked for being expensive and useless, but no informed critic suggests that the quality of the work being done, considered as basic science, is not of the highest order.

Distribution of qualified manpower
So much for the main features of Hagstrom's analysis of the scientific community. A number of qualifications and criticisms have been put. Chief among them is the one concerning the limited range of applicability of the scheme. It seems to apply only to those scientists who operate in an academic environment – a minority of all scientists. Let us now look at some figures to see how big a minority this is.

Data for people holding degree-level qualifications are summarised, rounded to the nearest thousand, in Table 1. They were obtained in the 1968 manpower survey which covered almost a quarter of a million qualified scientists and engineers (QSEs, in manpower jargon); the survey did not include small firms and some other categories. The total number estimated to be in employment, that is, the whole active stock of QSEs, is judged from population census results to have been one-third as much again.

Table 1. Persons holding degree or equivalent qualifications in science, technology and engineering in Great Britain, 1968 (thousands)

Industry	
manufacturing	111
construction	10
research associations	2
nationalised	22
Total in industry	145
Government	
government departments	15
research councils	3
atomic energy authority	5
local authorities	12
Total in government	35
Education	
universities	19
schools and further education	42
Total in education	61
Total in survey	241

Source: derived from *Statistics of Science and Technology 1970*, tables 48, 49 and 52.

Three main sectors of employment are distinguished: industry, government and education. Industry is easily the biggest, with manufacturing industry – engineering, chemical, electronics, oil refining, metal manufacture, aircraft, food, textiles and so on – taking well over a hundred thousand. 'Nationalised industry' includes transport corporations, electricity, gas and coal. The industrial research associations are organisations for co-operative research by members of an industry such as wool or linen or flour milling and baking. Member firms pay subscriptions which the government supplements proportionately. The arrangement is particularly suitable for industries in which there are many small firms which are not in a good position to do their own research.

These research associations are quite different, of course, from the research councils included in the government sector. Apart from awarding studentships and grants for research in universities, the councils also run research institutes. The largest of them by far is the Science Research Council, and in addition – excluding for the present purpose the Social Science Research Council – there are the Medical Research Council, the Agricultural Research Council and the Natural Environment Research Council.

Apart from some of the research councils' activities, such as the Science Research Council's support of high energy nuclear physics in laboratories based on large particle accelerators, most 'government' research is more like industrial than academic research in that it sets out to achieve practical goals set by social rather than scientific considerations. Government departments run research stations to tackle problems connected with defence, roads, building and so on. Whether such problems are tackled in a profit-making or a non-profit-making context is only a second-order difference. The first-order difference is that between internal and external criteria, between contributing to knowledge and solving practical problems. 'Government' research is overwhelmingly of the latter kind. For the former, it funnels taxpayers' money to universities.

Universities account for less than a third of the sixty thousand QSEs employed in the educational sector.[14] The remainder were in schools of various kinds, colleges of education and establishments of further (as distinct from higher) education – typically technical colleges, though the distinction between these and universities has become increasingly blurred with the evolution of colleges of advanced technology into technological universities and the creation of polytechnics.

From this quick survey it is possible to make a rough estimate of the proportion of QSEs to whom the information-recognition system of control is likely to apply. It is probably closer to 1 in 20 than to 1 in 10. Even if one estimates them as a proportion of those whose degree or equivalent qualification is in science rather than engineering or technology – about 104,000 – they are still very much in the minority.

Scientists in industry

How does sociological orthodoxy reconcile itself to the weight of the numbers of scientists who do not operate in an information-recognition exchange environment? By supposing that although they are not in fact in such an environment, they would like to be – in effect, that all scientists not working in basic research are relative failures, people whose ambitions have been frustrated. A scientist in industry, according to this view, finds himself in a conflict situation: his scientific values, the goals and norms of his profession, as acquired during education, conflict with the managerial values of his firm. He finds himself at the intersection between two 'cultures', that of science and that of management. His own goals are supposed to be those of basic science, the furtherance of knowledge – a very different matter from the goals of his firm, which are to achieve commercial or other practical objectives. He is unsure whose approval he ought most to seek: that of the world-wide community of specialists in his discipline or that of other members of the firm or organisation in which he works (cf. p. 82 above). And he is torn between two different kinds of incentive: recognition for his contributions to knowledge, or status and authority in the organisation, together with the higher income they imply.

There is a certain plausibility about this formulation, and to some extent this plausibility acts as a self-fulfilling prophecy. Scientists, and in particular science students, cannot help any more than other people being influenced to some extent by what others tell them they ought to want: prestige lies in the eyes of the beholders. On the other hand, it is clearly not true that all scientists in industry are frustrated academics. There are, thank goodness, plenty of people who find satisfaction in achieving practical results.

The facts of the situation are susceptible to study and surveys have been made of scientists in industry.[15] One such survey in Britain was reported by N. D. Ellis in a paper from which an extract is given in Appendix 6. Some of his results are summarised in Tables 2 and 3. Table 2 indicates how important the various factors listed are felt to be (how important do you feel

Table 2. Importance of various factors for work satisfaction

	UNIVERSITY		INDUSTRY	
	Scientists n = 50	*Technologists* n = 70	*Scientists* n = 118	*Technologists* n = 75
A. Salary	55	81	77	76
B. Quantity and quality of assisting personnel	79	65	78	63
C. The amount of free time available for private research	90	73	38	38
D. Opportunity for gaining experience in administration	20	32	63	55
E. Prestige of this department in the scientific world	60	67	43	51
F. Prospects for promotion up a research career ladder	60	70	75	91
G. Extent my qualifications and experience are fully utilised	81	95	94	85
H. The opportunity to pursue basic research in my field	90	70	33	49
I. Freedom to choose my own research projects	88	80	51	54
J. The degree of freedom I have to manage my own work	96	88	91	90
K. Opportunity to attend scientific/technical meetings/conferences	72	70	65	60
L. Opportunity to work with highly reputed technologists/scientists	63	72	44	50

Source: adapted from N. D. Ellis, *Technology and Society*, vol. 5 (1969), p. 33.

Table 3. Present level of satisfaction with various factors

	UNIVERSITY		INDUSTRY	
	Scientists n = 50	*Technologists* n = 70	*Scientists* n = 118	*Technologists* n = 75
A. Salary	31	40	50	52
B. Quantity and quality of assisting personnel	47	40	38	39
C. The amount of free time available for private research	74	66	49	65
D. Opportunity for gaining experience in administration	64	64	37	41
E. Prestige of this department in the scientific world	51	37	50	60
F. Prospects for promotion up a research career ladder	49	38	44	54
G. Extent my qualifications and experience are fully utilised	84	70	47	65
H. The opportunity to pursue basic research in my field	83	85	60	65
I. Freedom to choose my own research projects	96	85	49	56
J. The degree of freedom I have to manage my own work	93	93	63	79
K. Opportunity to attend scientific/technical meetings/conferences	60	45	51	67
L. Opportunity to work with highly reputed technologists/scientists	77	28	49	61

Source: as for Table 2.

that item is for your overall work satisfaction?), and Table 3 gives the present level of satisfaction for the same set of items (how satisfied are you at present with each condition?). Respondents were asked to indicate high, medium or low for each item and the ratings were summated to give indices in the range 0 to 100.

There is much food for thought in these tables. Note, for instance, the big difference between university and industry in Table 2, line C, which shows how much less highly free time for private research is valued in industry than in universities. Line H shows much the same for opportunity to pursue basic research. If these figures are to be believed, scientists and technologists in industry are not pining quite as longingly for the groves of Academe as the conflict theory would suggest.

Line D shows the opposite relationship for opportunity to gain administrative experience. University scientists score 20 in this line, easily the lowest score in either of the two tables; they want to have as little as possible to do with administrative chores. In industry, however, administration is not looked down on, presumably because it is seen as a route to positions of greater managerial responsibility. Moreover, it is intriguing to note that scientists in industry, quite unlike their university counterparts, view administration more favourably than technologists. The latter tend to bring with them from their training expertise which is more directly of use in the industrial context, and therefore feel more committed to their original specialism and less inclined to move out of it into more general managerial roles. It is consonant with this interpretation, that, as line F shows, they attach greater importance than do scientists to prospects for promotion up a research career ladder.

Many difficulties surround work on attitudes and motivation. Attitudes are fickle and climates of opinion change. There are certainly marked differences between countries. It is not easy to be sure how far respondents who express satisfaction are just putting a brave face on situations they have to accept anyway. Nevertheless, the data do strongly confirm that many QSEs in industry are well integrated into their work situations and well adjusted to their roles. They find positive satisfaction in them, and do not view themselves as having been thrust from the sweetness and light of universities into the outer darkness of the wicked world of industry and commerce.

Scientific manpower policy: problems
Let us now look at the kind of facts and considerations advanced by Hagstrom and Ellis in the context of scientific manpower policy. For this there is an obvious starting point: the important report on *The Flow into Employment of Scientists, Engineers and Technologists*, produced by a committee chaired by Professor M. M. Swann, then Principal of Edinburgh University, and published in 1968. Appendix 7 reprints from it the outline of its argument and the summary of its recommendations.

The key proposition from which its argument proceeds is the 'starving of industry and schools' of good graduates because of the undue attractiveness of fundamental research. Is there, then, despite all the criticisms of Hagstrom's thesis made above, all too much truth in what he says? Are students brainwashed by university education into seeking academic prestige through basic research? The Swann committee felt that the facts they examined gave grounds for strong suspicion that this was occurring to an undesirable extent. An aggravating factor, arising from the free international mobility of qualified manpower, was the notorious 'brain drain', documented in another report published in the same year by a working group on migration chaired by Dr F. E. Jones, Managing Director of Mullard Ltd and himself a member of the Swann committee. A Ph.D., the feeling ran, if not offered the opportunity to do in Britain the kind of research he wanted to do, had only to send a couple of letters to the United States of America to find himself wooed across the Atlantic by offers of jobs with tempting pay and conditions; and no way of plugging the drain seemed feasible without imposing unacceptable restrictions on the right of individuals to leave the country.

The Swann Report refers also to a third report published in 1968 – that on the flow of candidates in science and technology from schools into higher education. This inquiry was chaired by Dr F. S. Dainton, then Vice-Chancellor of Nottingham University and also a member of the Swann committee. Cynics said that the three reports together gave a complete picture of the typical scientific career: school to university first degree to Ph.D. and then down the brain drain. The Dainton Report documented the 'swing away from science', the decline in the proportion of sixth formers taking science subjects.[16] At the time, the threatened diminution in the supply of QSEs caused as much worry as the brain drain. Some critics, though, questioned whether the worries were consistent with each other. Did not the export of QSEs indicate that in some sense there must be a surplus at home? How badly could we need a greater supply when British industry was not absorbing those that were being produced? If the need were really desperate, surely firms would pay the price necessary to get the men they wanted. Was the shortage of QSEs perhaps more apparent than real?

The Swann committee's view was that industry was not attractive enough an employment prospect to get its fair share of the most able graduates. If studies such as that by Ellis are right in concluding that QSEs already in industry are not grossly dissatisfied with their lot, then the problem cannot be more than merely one of bridging a gap – of getting more realistic information about industrial careers to science students, of correcting unduly unfavourable impressions that they might have picked up and of helping them to make the necessary adjustments. This reading of the situation provides the rationale for the short courses that have since been mounted to infect postgraduate students with enthusiasm for the intrinsic interest and intellectual challenge of industrial problems.

Within three years of the appearance of the Swann Report, however, the situation already looked very different. The 'positively dangerous situation' of 1968 was transformed. The swing from science looked as though it was more or less halted, even in terms of the proportion of sixth-formers doing science, with their total numbers on the rise again. The brain drain had plugged itself, not so much because of any effective remedies being applied in Britain as because of changes in the American situation, notably the slackening of demand for QSEs due to contraction in the space programme. In Britain there was a substantial rise in unemployment, not only among QSEs but generally, and 'jobless graduates', previously almost a contradiction in terms, made ample newspaper copy. Graduates could no longer afford to turn up their noses at industry and commerce. Dr D. S. Davies of Imperial Chemical Industries, who had been a member of the Swann committee, was pleased to note a constructive result, in that good graduates must at last enter schools and the less prestigious small firms, as well as occupations where technical qualifications are not specifically required but where they give a useful background, in fields like merchant banking and local government.[17]

All this, with the possible partial exception of the slowing of the swing from science, was due not to the effect of exhortation in changing preferences but to the simple, stark realities of the employment market. The market is a harsh master, and also a fickle one. The transformation took only as long as a normal first degree course – a salutary reminder of the difficulty of planning in the educational and manpower spheres. In the face of essentially unpredictable vicissitudes of fortune, perhaps the safest prescription for education is to aim to develop maximum flexibility in students.

Scientific manpower policy: remedies

To remedy the situation as diagnosed, the Swann committee recommended educational reforms, without of course being so foolish as to suppose that these could in themselves suffice to work economic miracles. The general tenor was that education should be more closely matched to the needs of employers. For instance, universities were exhorted to consider drastic changes in the Ph.D. system (recommendation 3). A Ph.D. programme need not always be so much an apprenticeship for academic research; with a broadening of horizons, it could also include forms of postgraduate training more like apprenticeship in solving industrial problems. Would this debase the academic currency? Judging by the reactions of many university teachers, they had little doubt that it would. Certainly there was no widespread enthusiasm and Ph.D. programmes organised by universities jointly with industry have been established only on a small scale. The reluctance to arrange joint projects suggests a culture clash reminiscent of the model that Ellis attacks (p. 154). For instance, degrees are given to individuals but real life industrial problems tend to be tackled as collaborative efforts in which it may not be feasible or desirable to keep individual contributions distinct. As regards choice of problems, intellectual interest does not always go with practical usefulness, any more than the reverse correlation always holds. The time scales on which real problems have to be tackled do not necessarily match the three-year packets customary for academic Ph.D.s.

To improve the supply of science teachers for schools, the

Swann Report suggested establishing 'priority categories' for them (recommendation 8). Whether or not this might be a good idea in theory, in practice the principle of paying science teachers more than arts teachers is too hot for the school-teaching profession to handle, although of course hidden differentials can exist in the form of quicker promotion to posts of responsibility carrying higher salaries. The combination of science or technology with teacher training within first degree courses (recommendation 10) is practised on a limited scale and its extension is advocated,[18] but once again eagerness is not widespread, either among students or among university scientists – or, come to that, among university teachers of other subjects. Yet the idea seems eminently sensible from a vocational point of view. What explanation can there be for the lukewarm enthusiasm, other than that considerations of academic prestige conflict with the requirements of other forms of employment?

To meet longer-term needs, the Swann Report recommended 'making the first degree course in science, engineering and technology broad in character' (recommendation 13). Much has been said and written about broadening courses, a good deal less actually done.[19] Is it really a good idea?

Those who push with reforming zeal for broader courses would do well to distinguish more carefully than they commonly have in the past between different kinds of breadth. Combinations of one science subject with another science subject are usually easiest for a university to arrange, but it is often not easy to discern a positive purpose behind such combinations. Arrangements to add social themes to put scientific and technological subjects into perspective, or managerial topics with which they could usefully be combined in industrial employment, perhaps have more real point, even though they may at first sight look less likely in strictly academic terms. It was breadth of this kind that the Swann Report had in mind in suggesting 'relevant study in other fields such as economics, sociology, law, etc.'. Traditionalists, of course, suspect mixed courses of superficiality. But what a strange assumption they make when they suppose that only the weaker students are fit to tackle more than one subject! It is the brighter ones who have the mental agility to switch between different areas and styles of thought, based on different kinds of paradigms.

It might be asked whether broadening such as the Swann

Report suggested is appropriate only in circumstances in which graduates have to be coaxed and wheedled into accepting jobs in industry. Can it be dispensed with when the employment situation changes from a sellers' market for graduates into a buyers' market? In one sense, yes – but that would once again be bowing to academic pressures at the expense of helping students to make transitions in attitudes and styles of thought which many of them will have to make in any case when they leave university and take their first jobs. A majority of QSEs are employed in what the Swann Report calls 'an increasingly wide spectrum of occupations extending well beyond the traditionally vocational employment in these subjects and into fields outside technology as such'. It has long been the case that most QSEs are engaged not in research and development but in a variety of other functions for which 'managerial' is a convenient umbrella term. Only about a third of the QSEs in the industrial and government sectors are in the research and development function – roughly half the scientists and a quarter of the engineers. Moreover, these proportions are decreasing:[20] between 1965 and 1968 they dropped from 23·7 to 21·5 per cent for engineers, from 34·4 to 30·2 per cent for technologists and from 55·6 to 53·2 per cent for scientists.[21] Many QSEs who start in research move as they acquire experience into managerial roles. The difficulty science and technology graduates have been finding in recent years in getting the sort of first jobs traditionally regarded as appropriate for them is, in a sense, little more than a speeding up of what has long been a normal career progression. Research laboratories no longer act, to the extent that they used to, as points at which science and technology graduates can enter organisations and acclimatise to them. Does this not greatly strengthen the case for broader first degree courses?

6 Science and Wealth

The ethics of making science bear fruit

There is no getting away from it: wealth is what most people want from science. Idealists may thunder from the political Left about the alignment of politicians with big business and demand that science should be used for the benefit of people and not of profits, but the fact is that what the overwhelming majority of people want more than most other things is higher incomes. Economic growth, therefore, while not the sole and overriding objective of governments, has been a major one and is likely to remain so for some time to come. Hence the connection – or rather, the presumed connection – between science and wealth is in practice the most powerful justification for public support of science.

Having already met the antithesis between internal and external factors on a number of occasions in this book (pp. 22), we should have no difficulty in recognising that wealth and material progress constitute the most weighty of the external factors. In the last chapter, we considered what attitudes scientists in industry actually take on this matter of internal and external factors and the alleged conflict between them. Let us now look at the same matter in a rather different way and ask what attitudes they *should* take. What are the ethics of internalism and externalism?

It is hardly necessary to spell out the supposed moral virtues in internalism. The idea is firmly imbedded in much popular tradition and mythology that the virtuous, pure, noble, idealistic and selfless thing for a scientist to do is to devote himself to the 'disinterested' pursuit of knowledge (cf. pp. 81).

For the moral virtues of externalism, we can do no better than to go back nearly four centuries to the time when they were described by Francis Bacon with an eloquence that has not been equalled since. Bacon, significantly, was not primarily a scientist. As a lawyer and politician – he rose to become Lord

Chancellor in 1618 – he saw the potential benefits of science from the wider point of view of society, and he wrote about them with a happy gift for pithy aphorisms, 'a wonderful talent for packing thought close and rendering it portable', as Macaulay put it.

The main lines of his campaign, later developed at great length, are summarised in a concise and convenient form in a piece called *In Praise of Knowledge*,[1] which was written in 1592, when he was still quite a young man of 31. The pleasures of the intellect, he says in this short piece, are the greatest of all pleasures – but he goes on immediately to ask, are they to be 'only of delight, and not of discovery?' And so he very soon poses the momentous question: 'Is truth barren?' Just three short words, but they are pregnant with significance. Is knowledge to be only for its own sake, or can we use it as a help in doing things? 'Shall we not thereby be able to produce worthy effects, and to endow the life of man with infinite commodities?'

In the way Bacon goes on to condemn the kinds of learning current in his day, one can sense an undercurrent of moral indignation. They seem *wrong* – not necessarily and not only in the sense of being false but above all in the sense of being mis-directed. 'Are we the richer by one poor invention, by reason of all the learning that hath been this many hundred years?' 'All this is but a web of the wit, it can work nothing.' On the one hand, there is the academic philosophy of the universities, which consists of mere disputations about words; on the other hand, there is the alchemical tradition, which has at least some genuine contact with the physical world through experiment but is ruined nevertheless by the impostures that are rife in it and the obscurity in which its procedures are wrapped. 'The one never faileth to multiply words, and the other ever faileth to multiply gold.'

This futile ineffectiveness Bacon contrasts with the enormous impact made by three inventions of the late Middle Ages – printing, the use of gunpowder in firearms and the magnetic compass. Revolutions have been wrought by these in learning, in war and in navigation respectively. Yet they were 'but stumbled upon, and lighted on by chance'. How much more might not be achieved by a conscious and planned effort to generate more inventions! It would be quite possible, if only the world of learning could establish a 'happy match between the mind of man and the nature of things'. Through that kind of study, mankind could establish a genuine command over nature. 'The sovereignty of man lieth hid in knowledge' – in other words, unlocking the secrets of nature will give man mastery over it.[2]

Faced with such eloquent persuasiveness, can anyone still maintain that the 'disinterested' pursuit of knowledge for its own sake is the noble ideal for a scientist to pursue? Must one not accept that deliberately to seek applicable knowledge is by no means less high-minded but rather more so? Or even go further to say that the scientist who does not try to make himself useful is guilty of a sin, if not of commission then at least of omission?

The balance between internal and external factors

Having already stood back a few centuries in time, let us continue to use the perspective which history can give to put the question of internal and external influences in yet another way. What, in broad terms, determines the kinds of problems in which scientists interest themselves and the directions in which science advances? One view is that development is determined by the inherent logic of the subject-matter; the next steps to be taken are more or less fixed by the present state of knowledge within the area and, except for relatively minor and local accidents, perturbations and aberrations, the path of progress is the inexorable unfolding of a pre-existing pattern. The other view is that science is not so autonomous and scientists are not so shut off from other things going on around them. Economic and social factors external to science, including industrial, commercial, military and political ones, exercise the crucial effects in determining which topics become foci of interest and which lines of research attract most effort.

Historians have debated this issue at length. It is easy enough, with suitable selection of examples, to put a reasonably convincing case for either view. A useful collection of readings edited by Basalla[3] makes it easy for anyone with the interest to follow the main outlines of the controversy in the historical context.

One of the key contributions was made by the American scholar R. K. Merton.[4] Taking a sociologist's look at science, Merton studied the topics chosen by scientists in seventeenth-century England by analysing the minutes of the newly founded Royal Society for the years 1661, 1662, 1686 and 1687. Well over half of the researches discussed at meetings during these years, he found, were related directly or indirectly to socio-economic needs, principally in the fields of marine transport, mining and military problems.

The interpretation of these facts is of more than purely historical interest. While there is no need to lapse into the 'vulgar materialism' of supposing that socio-economic factors account exhaustively for the whole complex of scientific activity, it is quite clear that the dominant themes of science in seventeenth-century England were determined to an appreciable extent by the social conditions of the time. Merton argues that it would be a mistake to suppose that the only channel for this influence was the deliberately utilitarian motivation of some scientists. There is also a less direct relationship between science and social needs in that 'certain problems and materials for their solution come to the attention of scientists although they need not be cognisant of the practical exigencies from which they derive'. Indeed, analysing motives may give a quite misleading picture of the modes by which socio-economic factors exert influences on science. The motives of scientists may range from personal ambition to a wholly disinterested desire to know, but that does not alter the demonstrable fact that the subjects that seventeenth-century scientists chose to work on were largely in areas which seemed relevant to the major practical problems of the time. Merton concludes that it can be said, with suitable provisos, that 'necessity is the (foster) mother of invention and the grandparent of scientific advance'.[5]

Is the same general conclusion not still valid today, with military demands remaining a major factor, however unfortunate that may be (cf. p. 15), with 'mining' widened to include general industrial and commercial needs, and with the place formerly held by marine transport now taken by aerospace? Seen in this light, there is nothing very new about the pressures on research and development generated by the space race and the military-industrial complex, which are so bitterly attacked from some quarters. There is impressive historical continuity about the way in which factors like these have guided and stimulated research effort. 'If society has a technical need, that helps science forward more than ten universities', wrote Engels.[6]

The important thing, surely, is that the technical needs should be continually revised so as to keep them in line with current social priorities. Of course people disagree about these priorities, and denounce those of which they disapprove for the way they 'distort' the research effort; but then, determining social priorities is the sort of political issue on which, by definition (p. 50), unanimity is not to be expected.

It is not difficult to see the broad currents identified by historical analysis operating in current science policy. Consider the *Framework for Government Research and Development* published by the British government in 1972. It announced the decision to transfer some of the funds of the Agricultural, Medical and Natural Environment Research Councils (cf. p. 87) to appropriate government departments over the period 1973–6. This transfer, involving £20 million (at 1971–2 prices) of the total annual budget for the three research councils of £56 million,[7] was unpopular with much of the scientific community but was justified in the eyes of users of research by the hope that it would strengthen the coupling between research and practical needs.

To take an optimistic view, effective coupling should be helped also by the explicit formulation and official acceptance of the 'customer/contractor principle'. This principle is supposed to guide the organisation of applied research and development commissioned by government departments. 'The departmental "customers" must work in partnership with their research and development "contractors", whether inside or outside the department. Responsibilities are then clear. Departments, as customers, define their requirements; contractors advise on the feasibility of meeting them and undertake the work; and the arrangements between them must be such as to ensure that the objectives remain attainable within reasonable cost. This is the customer/contractor approach.'[8] In practice, it meant relatively little change in those departments which spent most on research – the Ministry of Defence (£330 million in 1972–3) and the Department of Trade and Industry (£109 million on civil aerospace, £43 million on reactor and other nuclear research, £26 million on other industrial research). The Department of the Environment, spending £17 million on research connected with planning, transport, construction, environmental pollution and resources, was a long way behind, in the same league as Agriculture, Fisheries and Food (£15 million) and Health and Social Security (£13 million).[9]

Push and pull models of innovation
Let us now once again transpose slightly the terms in which the internal/external problem is formulated and focus attention on the process of technological innovation. If one asks what are the ways in which science relates to innovations in productive processes, one soon sees that there are two possibilities, which correspond roughly with the internal and external views and which can conveniently be labelled 'push' and 'pull' respectively. According to the first, the impetus comes from discoveries: research yields a lot of results, some of which turn out to be useful and are later applied. It is virtually a matter of 'knowledge pressure' pushing its way into application. According to the second view, however, the impetus comes from certain needs or demands which generate the work needed to meet them: it is the desired end to be attained which 'pulls' the process. Changes in the social environment 'cause' inventions to arise just as much as inventions cause changes in the environment.[10]

Both these views are currently held. Explicitly or implicitly, they underlie a good deal of what is said and written about innovation. Amongst scientists the first is more prevalent, for reasons which are not difficult to see: it is the 'scientist's eye view' of the process. Thus P. M. S. Blackett, winner of the Nobel Prize for physics and President of the Royal Society, wrote: 'In a simplified schematic form, successful technological innovation can be envisaged as consisting of a sequence of related steps: pure science, applied science, invention, development, prototype construction, production, marketing, sales and profit. Clearly the first steps . . . cost money and only the later stages . . . make money.'[11] The matter looks different, however, as seen from the United States Department of Commerce by J. H. Hollomon: 'The sequence – perceived need, invention, innovation (limited by political, social or economic forces) and diffusion or adaptation (determined by the organisational character and incentives of industry) – is the one most often met in the regular civilian economy'[12] (see Appendix 11).

A 'push' view can be used to argue for the virtues of undirected research. It can be said even of the most esoteric academic research, far removed from any practical application at present conceivable, that, quite apart from its 'cultural' value in enlarging the intellectual space in which we live, 'you never know when it will turn out to be useful'. Romantic stories are told to illustrate the possibilities of unexpected application. One of the best is about Faraday being asked by a statesman what was the use of his work on electricity. There are two versions of the reply. One is 'Some day you'll tax it.' The other, even more open-ended, is 'What is the use of a baby?'

On the 'pull' view, this line of argument for supporting research irrespective of the likelihood of application becomes much weaker. It does not seem very wrong to nudge research more or less firmly into those directions which seem more likely than others to yield practical benefits in the short or medium term. Indeed, allowing research to proceed without at least gentle guidance of this kind might even impede progress by diverting scarce resources of able men and costly materials into unfruitful lines.

Both push and pull type views have appeared in official or semi-official science policy thinking. To illustrate the former, I have chosen a proposed method of estimating the economic benefits from basic research; for the latter, a government publication on measures that could be taken to promote effective technological innovation.

The Byatt–Cohen method
A suggestion for a way to quantify economic benefits from scientific research was published in 1969 by I. C. R. Byatt, Senior Economic Adviser in the Department of Education and Science, and A. V. Cohen, Scientific Secretary to the Council for Scientific Policy. One of the Council's major preoccupations was to act as a spokesman for science and to justify expenditure on research. Byatt and Cohen did not, of course, suppose that the economic benefits are the only ones worth considering. They gave a careful list of the various benefits, classing them under four main types as follows.

(i) Manpower benefits, namely the output from higher education of graduates and postgraduates taught by those active in research.
(ii) The benefits of applied research, also called 'mission-

oriented' research to indicate work done where the area of application is known.
(iii) The benefits resulting from the application in the economy of fundamental discoveries made in basic or 'curiosity-oriented' research.
(iv) The cultural values of scientific research.

It was the benefits accruing under (iii) that the Byatt–Cohen method was intended to measure – not the benefits under (ii), which are relatively obvious, however hard it may be to calculate their precise magnitudes, but those resulting from 'pure' research. The terms mission-oriented and curiosity-oriented are expressive ones, somewhat more explicit than the commoner but vague and subjective usage of distinguishing applied from pure, basic or fundamental work. Mission-oriented research is 'research in fields whose application is evident, for example work in agriculture and medicine'. Ideas discovered in curiosity-oriented research, by contrast, 'tend to arise unpredictably and, when applied, to give rise to big industries, or to complete re-orientation of existing industries, perhaps several decades later, though it is often maintained that this period is shortening' (cf. p. 23).

The line of argument on which Byatt and Cohen based their proposal runs as follows. Occasionally a discovery made in curiosity-oriented research finds application and is therefore followed by expenditure on applied research, development and investment in production facilities. If things go well, there is eventually income from sales of the product, and perhaps receipts from licensing agreements, to set against these expenditures. All these cash flows can be discounted to a common year to give the net benefit from the innovation, which may of course be negative, but will be positive in cases of success.

Now imagine, said Byatt and Cohen, what would have happened if, because of less effort in curiosity-oriented research, the crucial discovery had not been made until somewhat later. Usually the net benefit would then be lower, and the difference in benefits would represent the value of the increment in resources devoted to curiosity-oriented research. The whole method depends, then, on estimating the effects of a 'notional marginal delay' in a discovery.

The extracts from Byatt and Cohen's paper given in Appendix

8 should help to make the argument clear. Four points of explanation and elucidation should perhaps be added. First, there is no great mystery about what it means to discount cash flows to a common year. Discounted cash flow or DCF technique in essence just means applying the principle of compound interest to find what is the value in any given year of a given sum a given number of years earlier or later. The DCF formula shows that the present value of £100 to be received in n years' time, with an annual discount rate of r, is

$$\frac{100}{(1+r)^n}$$

It adds a time dimension to money values, and a few illustrative figures show how important it is not to neglect this. At a discount rate of 8 per cent a year, £100 which will be received in 10 years' time is worth only £46 today; if it is not to be received until 30 years have elapsed, its present value is reduced to £10. (This factor is, of course, independent of inflation and additional to it.)

Second, it is an important fact about technological innovation that the total cost of bringing it about is usually very much greater than that of the research that was necessary for it. It is sometimes said that the research cost is typically only one-tenth of the total cost, and although the variation between individual cases and the difficulties of definition and demarcation are so great that this cannot properly be regarded as anything more than the very roughest rule of thumb, it does serve to bring home to researchers that a bit of clever work on their part is by no means all that is necessary. Setting up large-scale production and launching the product on the market require heavy investment. To apply the Byatt-Cohen method, therefore, would necessitate estimating the difference between two large magnitudes – never a comfortable situation to be in.

Third, Byatt and Cohen clearly had in mind a push model of innovation. It was a discovery which they saw as the initiating impulse that sets in motion a chain of events culminating in a process used in industry to manufacture goods and create wealth. Fourth, moreover, their model embodies a particular view of the relation between basic science and technology. The technological know-how on which the industrial process is based was seen as an application of a discovery made in curiosity-oriented research. 'Science discovers, technology applies' is the assumption that underlies their approach.

Test of the Byatt–Cohen method
In a paper published in 1970, reprinted here in Appendix 9, Gibbons and others scrutinise in particular this last assumption and find it wanting. After examining a substantial number of innovations, they conclude that it is really quite hard to find a case that fits the Byatt–Cohen mould. Only in very rare instances is it possible to pinpoint a specific curiosity-oriented discovery from which a wealth-producing application is derived. The only clear twentieth-century cases of this kind that could be identified are nuclear power and silicones. Even here, it was not possible to apply the Byatt–Cohen method because the timing was influenced critically by the Second World War. We are back to the sad but apparently unavoidable fact that there seems to be nothing like a major war for stimulating technological development (cf. p. 15). The development of nuclear power clearly owes a lot to the intense cost-no-object effort put into the production of the first atom bombs. Silicones had been synthesised in the laboratory in the first decade of this century but were not produced commercially until 1943.

What many people find startling about these conclusions is the assertion that hardly any industrial processes are based on discoveries made in curiosity-oriented science. Can it really be true that all the common talk about the rise to economic importance of 'science-based industry' is founded on a misconception? Anybody who has even a modest acquaintance with science and technology would do well to think hard about any examples he happens to know of or can find out about. Whatever the conclusion he reaches, the effort is likely to have a salutary effect in sharpening the woolly notions that abound on the question of the relations between science on the one hand and technology on the other. For anyone who is going to embark on the exercise, there are three caveats to bear in mind.

First, it is necessary to be strict about the definition of 'curiosity-oriented' work as work in fields where no application is evident. With a few notable exceptions such as high energy nuclear physics (p. 28), some kind of application is conceivable for most areas of scientific research. This may sound suspiciously like academic hair-splitting, but it is not only that, since it is relevant to the real policy problem of whether and how much to fund research regardless of the likelihood of economic or other social benefits. It makes the difference between saying to the applicant for research funds, 'Good luck to you, however useless your work may seem', or 'Give me some justification in terms of practical usefulness, however vague and distant'. In short, it means distinguishing between 'pure' and 'strategic' research.

Second, scientists themselves are frequently and quite genuinely unaware how 'impure' their research really is. They may think they are following the dictates of curiosity only, when the detached observer has no difficulty in identifying external influences which may well have helped to nudge them into the fields they are active in. As Merton argued for the seventeenth century (p. 100), deliberately utilitarian motivation is by no means the only way in which social conditions can affect scientific activity. Thus physicists are wont to expostulate indignantly that there is a very clear instance of curiosity-oriented research paying off in the way that A. H. Wilson's work in 1931–2 on conduction in semiconductors led to transistors; but a close look at the history of this case shows how well developed was the use of crystals as rectifiers in detecting radio waves.[13] Indeed, the 'crystal set' was the popular domestic radio of the 1920s. Whether or not Wilson realised the potential commercial importance of his work, he can hardly have been unaware that there were problems in explaining certain 'anomalous' effects in crystals.

Third, nobody argues that curiosity-oriented research is useless in economic terms. The contention is only that the mechanism by which it becomes useful is hardly ever as simple as that postulated in the Byatt–Cohen model. The science-technology relationship is less direct, and there is a great deal more to technology than just applying the results of basic research. It is not often that a curiosity-oriented discovery is applied to give a new process or product. Much more frequently, science helps by improving processes or products that already exist. There is no doubt that the increased understanding afforded by Wilson's theory of semiconduction helped to pave the way from crystal rectifier to transistor. As another example, take the case of the float glass process, described in Appendix 9. In this process, flat glass is made by floating it on a bed of molten tin. This can hardly be described as based on a discovery of curiosity-oriented research, but rather on a technological discovery. However, science helped to trace the cause of difficulties that arose. The diagnosis of a surface 'bloom' on the glass as due to dissolved tin clearly depended on concepts and techniques derived from basic research, some of which may have been curiosity-oriented. A similar dependence exists in many, if not most industrial processes.

Concepts and techniques become available to industry for the solution of its problems mostly by the recruitment of qualified manpower. To an appreciable extent, therefore, it seems that the economic justification for academic research comes not so much from category (iii) on p. 104, the application of any discoveries that might arise, as from category (i), manpower benefits – the output of graduates from institutions of higher education in which research is carried out. What industry needs most, it has been said, is not science but scientists.[14] This does not, however, mean that contributing manpower to be recruited and discoveries to be used are the only routes through which academic science can come to be involved in technical problem-solving in the context of industrial innovation. Other forms of coupling between the scientific and technological communities exist. Employment of university scientists as consultants to industry sometimes plays an important part. Universities may also undertake sponsored research, or provide specialist facilities, or give assistance and advice in a variety of more or less informal ways.

Commercial awareness
Consider now another government publication which tackles the problem of innovation with a different emphasis. The Central Advisory Council for Science and Technology published in 1968 a report on *Technological Innovation in Britain* which brought a new degree of commercial awareness to government level discussion of science policy. Newly established in 1967, the Council included, besides distinguished scientists, some high level industrialists and an economist. B. R. Williams, Professor of Economics at Manchester University, served on the Council during the first year of its existence.

Williams had done some work on international comparisons of expenditure on research and development (R and D) which showed a notable lack of correlation between research effort and economic growth.[15] On the macro scale, at national level, where individual cases of good and bad luck should cancel each other out, it did not seem as though high R and D effort leads to faster increase in gross national product. In Figure 6, R and D expenditures as a percentage of GNP for 1950 to 1959

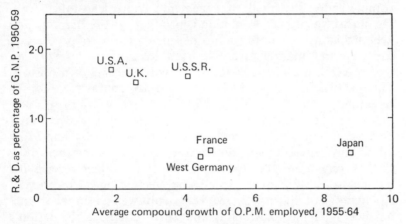

Fig. 6. Research and development expenditure, as a percentage of gross national product, plotted against average compound growth of output per man employed.

Source: Williams, *Technology, Investment and Growth*, p. 57.

are plotted against growth rates of output per man employed for 1955 to 1964, the five-year lag being introduced to allow time for the research effort to pay off. It is very clear that the countries which had put most into research were not those which had achieved the highest growth rates.

The issue here is not mainly concerned with the effect of basic science on technology, as it was in considering the Byatt–Cohen method, because the expenditures plotted are for total R and D, in which applied and development work preponderates heavily over basic research (p. 152). The question at issue is more the relation between technology and wealth than that between science and technology. What Williams' data seem to show is that even applied research and development do not necessarily lead to economic growth. 'Turning science more applied' is not enough; to create wealth one has to pay due attention to market and commercial factors beyond the realm of science and technology as such.

This philosophy stands out in the Central Advisory Council's report, as is evident in the extract from it which forms Appendix 10. Five factors are listed as being of particular importance for success in technological innovation: direct linkage of R and D with the other activities of the organisation; planning of innovation in the light of sophisticated market research; commercially dedicated management; short lead-times from the start of a project to getting a product on the market; and keeping launching costs in proper proportion to production capacity and market size. No sign here of any suggestion that research is a virtuous act which will bring its own reward! The exhortation is to feel dedicated to commerce, not to research.

Indeed, the report virtually throws a wet blanket over R and D by pointing out how well some countries do by importing know-how instead of spending heavily on generating their own. If they pay more for licences bought than they receive for licences sold, they are in an adverse 'technological balance of payments', but this may have a very healthy effect on the overall balance of payments, which is what really counts in the nation's trading position. This makes sound economic sense. Technologists, however, can be expected to resist the invitation to go negative on the technological balance of payments. They tend not to be immune from what is sometimes referred to as the 'NIH factor', the 'not invented here factor'. For obvious reasons, in which prestige plays a large part, they find making their own inventions a more attractive proposition than developing or adapting ideas imported from elsewhere. Technology, like science, has an internal impetus of its own, and this makes necessary the warnings about lead-times and market sizes, which refer in particular to technologically glamorous areas like advanced nuclear reactors and fast aircraft. Very large sums of government money have gone into both these fields (cf. p. 101), amid doubts whether the benefit to society is really commensurate with the cost to the taxpayer.

The purpose of direct linkage of R and D with manufacturing, marketing and financial activities is to try to ensure that the R and D staff invent and develop things that can be used. Without proper integration into the general functioning of the organisation, there may be a danger that they devote their efforts to technically ingenious ideas which cannot be put to practical use for production or commercial reasons. To bring out this point, it may be helpful to note the distinction which economists now commonly make between invention and innovation. Innovation is much more than just invention; it means actually establishing a process in use for producing goods or services. Most inventions are never turned into innovations. Hundreds of thousands of them never get far beyond a gleam in an inventor's eye, or lie in the Patents Office as patents which have never been exploited. This is not a scandal of creativity allowed to run to waste – it just means that no adequate market or social need has been seen which they could fill. It is always possible, of course, that changes in market or social conditions may bring them into use. But an efficient R and D manager will aim at products or processes that his organisation can use in the foreseeable future, not ones that merely add to mankind's storehouse of creative but unused invention.

For success in innovation, therefore, one would expect that sheer technical virtuosity will not be more important than clear definition and thorough understanding of user needs, acquired by careful study of the market and perhaps even direct collaboration with potential customers, and followed up, when there is a product to market, by adequate publicity, user education and continuing effort to anticipate customer problems. Close study of the process of industrial innovation in fact confirms the importance for success of attention to user needs and marketing.[16]

In effect, the message to R and D teams is that they should avoid aiming for 'internal' recognition, eschew the NIH factor and deliberately try to make themselves more amenable to need pull. Further, in line with this general aim, the Central Advisory Council's report urges that more men of the highest ability, including some of those with technical backgrounds, should be deployed in phases of innovation other than R and D, such as production and marketing. Of course this raises again all the problems of prestige and self-esteem that were brought up in the last chapter. Is a scientist or technologist demeaning himself by adapting other people's inventions instead of making his own, or by forsaking the laboratory to enter the marketing or commercial functions? If he lets himself be need-pulled instead of exerting discovery push, is he becoming just as much a victim of manipulative marketing as the unsuspecting customer?

Or is he, rather, putting his technical expertise at the service of society without presuming to tell society what it ought to want? These are questions which everyone must, in the last analysis, answer for himself. What we can do here in conclusion, though, is to get some idea of the relative importance of the parts actually played by push and pull in bringing about innovation in the real world in modern times.

Relative importance of push and pull
Some estimate of this can be formed from a study of eighty-four successful innovations by Langrish and others, reported in *Wealth from Knowledge*. The sample consisted of innovations which won Queen's Awards to Industry in the first two years, 1966 and 1967, of the operation of the scheme, which is designed to reward good performance in exports, in technological innovation, or both. In the criteria for innovation awards, a clear distinction is drawn between invention and innovation. 'We do not recommend', said the committee which drew up the scheme, 'that inventors or invention should be recognised as such. The scheme should concentrate on the practical application in industry of advanced technology whether in the form of processes or products.' The felt need, clearly, was to encourage effective innovation, not abortive invention, and the innovations singled out for awards are possibly representative of the kind that contribute to national wealth.

An excerpt from *Wealth from Knowledge* forms Appendix 11. Push and pull types of innovation were each subdivided into two categories, the push type according to whether the discovery that gave the impetus was a scientific or a technological one, the pull type according to whether the need was a customer need or some other need felt by management, such as reduction of manufacturing costs or avoidance of a take-over. Even with this refinement, many of the innovations did not fall neatly and unambiguously into any of the four conceptual models provided by the classification. The fit with observed facts could be improved somewhat, however, by treating the models as complementary rather than mutually exclusive, so that a given innovation can be regarded as brought about, from the point of view of the award-winning organisation, by a combination of two types of mechanism. Of the eighty-four cases, thirty-five seemed to be best described as dual cases of this kind, mostly with both push and pull operating: what Dr Dolittle fans might want to call the 'pushmi-pullyu model' of innovation.

Counting all the occurrences in single and dual types, it was found that need pull was important just twice as often as discovery push. There was some indication, however, that where large changes in the technology were involved, the incidence of discovery push was somewhat greater.

We should not leave this subject without a warning against over-enthusiasm about models, and linear models in particular. The overwhelming fact about innovation is the formidable complexity and variability of the circumstances surrounding it. Those without direct experience of this complexity can sample it vicariously in the case studies which form the bulk of *Wealth from Knowledge*. Squeezing it into conceptual moulds is like trying to fit a giant amoeba on a Procrustean bed. The virtue of models is, of course, that they bring some order into the chaos of raw observations and achieve some measure of intellectual economy. But does this demand too high a price in terms of fidelity to the real world? There are so many historical strands converging on each innovation, so many events affecting each other, that linear schemes, push and pull, separately or together, do not seem adequate to cope. We really need schemes with multiple interconnections, feedback loops and so on models on an altogether more sophisticated level.

7 The Future: Is Technological Optimism Justified?

Critical science

What lies ahead? In what ways will the activities of science and technology, and the roles they play in society, come to differ from what they have been in the past? Forecasting is difficult – especially when it concerns the future, as the joke goes. But since it is all too clear that the future is full of questions and problems, and undeniable that science and technology are major agents of change, it is important to try to look ahead.

One scholar who has thought seriously about this is J. R. Ravetz, from whose book, *Scientific Knowledge and its Social Problems*, Appendix 12 is taken. It puts forward the idea that a part of science in the future should be 'critical science'. The word 'critical' is used here in a sense other than, or at least wider than, the one that is attributed to it by a Popperian methodology of natural science in which ruthless criticism is the second pillar of scientific wisdom (p. 60). Rather, it is used in the sense of socially critical: humanitarian, setting out to analyse damage inflicted by technology on man and nature and to correct the poisoning of the environment.

Ravetz writes with a streak of left-wing idealism reminiscent of Bernal. Some may feel that he does less than justice to the beneficence of science in the past, and to the part played by bureaucracy in that beneficence. Established administrative machinery does not always block progress; it can also be the means by which progress is achieved. Furthermore, it is a highly dubious assumption to suppose that science need only be freed from the repressive yoke of the military-industrial complex to solve the problems of our environment. Even if military and civil industry were wished away, the tension between internal and external factors would remain, with a revised set of external criteria. There is no reason to suppose that the linkage of basic science to environmentally directed technology would be less tenuous than that to wealth-creating technology (p. 106); in any case, these two kinds of technology overlap to a large extent.

Ravetz is surely right, however, to emphasise the inevitability of the political dimension. Science and technology already play important parts in public affairs, and there can be no return from that position, though there is room for disagreement concerning the appropriate style of politics.

How can we set about defining somewhat more specifically what the aims of critical science should be? 'Poisoning the environment' is vague and subjective, especially when the word 'environment' is used in the all-embracing sense it has recently acquired – meaning all, and more than all, that was understood by older and now less fashionable expressions like 'quality of life'. On such issues there can be no unanimity (p. 50). Yet they include thorny problems of the most sweeping and crucial significance. Plenty of people are convinced that it is not just the quality but the very survival of human civilisation that is at stake, and scientists, especially life scientists, have been prominent among those who have sounded danger signals.[1] It is possible to see so many difficulties ahead: even excluding the threat of nuclear, chemical or biological war, there remain the population explosion, escalating pollution, imminent exhaustion of non-renewable resources, increasing inequalities in the distribution of wealth between countries and within them, psychological stresses on individuals, extinction of living species, loss of soil fertility. . . . The list could be extended quite easily.

Limits to growth?

Because there are so many variables and interrelationships to follow simultaneously, some workers have constructed models in which they can be handled by computers. In one such study, carried out by a team at the Massachusetts Institute of Technology, a world model was set up containing five main kinds of quantities and the rates of the processes that relate them to each other, the five quantities being population, capital (industrial, service and agricultural), land (agricultural and urban), non-renewable resources, and pollution. With no major change in the physical, economic and social relationships that have governed world development in recent history, the model shows collapse by the middle of the twenty-first century because of non-renewable resource depletion.[2] With more generous assumptions about world reserves of natural resources, collapse occurs because of a sharp rise in pollution.[3] To get a stable system, it was necessary to stop the growth of population and of industrial capital in the near future, together with other measures – substantial resource recycling and pollution control, a swing in social preferences from material goods to services, restoration of eroded and infertile soil and increased durability of industrial capital.[4]

Such findings, if valid, would be political dynamite. Limits on population and economic growth are, to put it mildly, contro-versial issues, and there has been no lack of critics to point out the many uncertainties which remain regarding the com-puter model, the assumptions that underlie its structure and the data that were used as inputs to it. One of the major assumptions is of particular interest here. It is that regarding the ability of technology to find ways of getting around the apparent limits to growth.

The Massachusetts team satisfied themselves that tech-nological solutions would not suffice. 'The hopes of the technological optimists centre on the ability of technology to remove or extend the limits to growth of population and capital. We have shown that in the world model the application of technology to apparent problems of resource depletion or pollution or food shortage has no impact on the *essential* problem, which is exponential growth in a finite and complex system. Our attempts to use even the most optimistic estimates of the benefits of technology in the model did not prevent the ultimate decline of population and industry, and in fact did not in any case postpone the collapse beyond the year 2100.'[5] Therefore, concluded the Massachusetts workers, it is necessary to resort to non-technological policies: to put a brake on the growth of population and industry.

Of course, no one in his senses denies that growth cannot continue indefinitely with anything like the annual percentage increases that have taken place recently. No fact about exponen-tial growth could be more obvious. We noted it earlier (p. 38) in the more limited context of the growth of science as measured by manpower or publications – Price's 'second law'. The real question is what kind of mechanism it will be that applies the brake and how unpleasant it will be. It could be global famine, pestilence or war; but that sort of outcome we would like to avoid.

Some people maintain that the pace of growth is bound to slacken in any case due to the operation of factors already at work, principally the economic forces of the market place. Extrapolating into the future, even if the annual decreases in the percentage growth rate are small, they lead over a long period to figures very different from those given by true exponential growth, in which the percentage growth rate remains constant by definition. It is claimed that some large-scale economic indicators, such as energy consumption per head, have indeed shown this kind of 'tapered growth' in the past.[6] But it might be somewhat misleading to conclude from such data that there is some 'automatic' regulation at work – that growth is 'self-limiting'. The implication could be that we can afford just to sit back and let events take their course. As Ravetz says, technology cannot be held to be self-correcting. Adjustments may appear to be automatic if viewed on a gigantic scale, but on a human scale one cannot assume that technologies to provide the necessary capabilities will rain like manna from heaven and that the accompanying social adjustments within and between nations will be looked after by some superior wisdom. They will not: human effort will be required to bring them about. As best we can, we will have to decide what new technologies to try to develop, what existing ones to adapt and adopt, what political and administrative machinery to use. In particular, human foresight is needed because of the delays with which natural regulatory mechanisms operate. Time lags in the feedback loops could lead to disastrous overshoots.

For population limitation, technical means have become available but it is doubtful whether they are coming into use fast enough. Birth control has hardly begun to make an impact on, say, the population increase of a million a month in India. Some less developed countries seem to consider themselves underpopulated and are therefore not likely to accept population stabilisation as an acceptable political objective at present (p. 18).

The controversy over economic growth centres on the advanced countries, since the need for it in less developed areas of the world can hardly be contested while a population of some 1,700 million live in countries with gross national products per head of less than 100 dollars a year.[7] Should people in Western Europe and North America abate their demands for ever-increasing material standards of living? Profound changes in value systems would be involved, and there are doubts how far they are necessary or desirable, expressed by economists[8] as well as scientists, including the ex-editor of *Nature*, John Maddox.[9] Economic expansion, it is argued, is more a cure for present ills than a cause of them, and more an insurance for the future than a threat to it. Cutbacks would quickly exacerbate unemployment problems. Those who call for zero growth seem to ignore the fact that yardsticks of economic expansion like gross national products include services as well as output of factory-made goods: provision for health and education as well as production of chemicals and cars. They also include the costs of pollution control. Since technical means are available in most cases, pollution charges could, so the argument runs, effect the reallocation of resources necessary to combat pollution problems. Money for environmental improvement seems more likely to be found out of expanding gross national products than out of static ones.[10]

As for depletion of natural resources, the world has survived plenty of alarms in the past. In the nineteenth century there were warnings that industrial expansion might be restricted by exhaustion of coal supplies and that diminishing supplies of fixed nitrogen might mean starvation; but alternative fuels have become available, and chemistry has made it possible to fix atmospheric nitrogen.[11] Nuclear energy has now released us from dependence on fossil fuels; fast breeder reactors are likely to extend our stocks of fissionable fuels, and in the more distant future there is the prospect of vast quantities of energy from nuclear fusion. Supplies of many metals are predicted to run out within a few decades; even at current rates of consumption, without allowing for growth, known global reserves are said to be insufficient to last out the twentieth century in the cases of gold, mercury, silver, tin, zinc and lead, and the situation is not much better for copper and tungsten.[12] But known global reserves have a habit of expanding as exploration continues and as technology makes it feasible to use deposits lower in quality or more difficult of access. Besides, when shortages loom the price mechanism can be expected to stimulate technology to provide substitute products and processes.

In the light of considerations like these, is it justifiable to adopt an attitude of technological optimism? Probably that is the most reasonable line to take, as long as the optimism is cautious and strictly conditional. Events are likely to take a course somewhere between the extremes of gloom and confidence. If we try hard, we can more effectively combine technology, existing and new, with social requirements which themselves will change, at least in their ranking of priorities. In the long term, certain trends in technological innovation can be predicted with some confidence: to conserve or recycle non-renewable resources, to produce less pollution, to focus less on labour saving and to be less capital-intensive, to be directed more to the service sector and less to material goods. These trends will be gradual rather than sudden. Further extensions of technological capabilities will be called for but the most pressing need will be for much more concerted and sophisticated attempts to foresee and estimate the less direct consequences of changes that are contemplated, so as to work out in more detail what are likely to prove the best courses of action. The problems of the future will be problems more of technology assessment than of technology itself.

References

Chapter 1
1. Galileo, letter to Belisario Vinta; quoted in S. Drake, *Discoveries and Opinions of Galileo* (Doubleday Anchor Books, New York 1957), p. 63.
2. J. B. Conant, *Modern Science and Modern Man* (Columbia University Press, New York 1952), pp. 8–9.
3. C. P. Snow, in M. Goldsmith and A. Mackay (eds), *The Science of Science* (Penguin Books, Harmondsworth 1966), p. 29.
4. Department of Education and Science, and Ministry of Technology, *Statistics of Science and Technology 1970* (HMSO, London 1970), Table 2.
5. *Report from Iron Mountain on the Possibility and Desirability of Peace*, introduced by L. C. Lewin (Penguin Books, Harmondsworth 1968), pp. 84–6.
6. *Pollution: Nuisance or Nemesis?* (HMSO, London 1972), para. 138. This report of a working party chaired by Sir Eric Ashby gives a balanced account.
7. Roger Williams, *Politics and Technology* (Macmillan, London 1971), p. 58.
8. J. R. Ravetz draws attention to this play in *Scientific Knowledge and its Social Problems* (Oxford University Press 1971), p. 428 n.
9. H. Ibsen, *An Enemy of the People*, Act II.
10. ibid., Act IV.
11. *Pollution: Nuisance or Nemesis?*, para. 57. See also H. Rothman, *Murderous Providence* (Rupert Hart-Davis, London 1972), pp. 98–112, 209–11, 230–2.
12. P. M. S. Blackett in *The Science of Science*, p. 49. Also in his introduction to Graham Jones, *The Role of Science and Technology in Developing Countries* (Oxford University Press 1971), p. xii.
13. M. Gibbons in J. Knapp, M. Swanton and F. R. Jevons (eds), *University Perspectives* (Manchester University Press 1970), p. 136.
14. M. Polanyi, *Minerva* vol. 1 (1962), p. 54.
15. D. S. Greenberg, *The Politics of American Science* (Penguin Books, Harmondsworth 1969), p. 19.
16. H. Rose and S. Rose, *Science and Society* (Penguin Books, Harmondsworth 1970), p. 233.
17. Greenberg, *The Politics of American Science*, p. 355.
18. F. R. Jevons, 'How Valuable is Biochemistry?', *New Scientist* (3 Feb. 1966).
19. J. K. Galbraith, *The New Industrial State* (Penguin Books, Harmondsworth 1969), p. 71.

20. J. Langrish, M. Gibbons, W. G. Evans and F. R. Jevons, *Wealth from Knowledge* (Macmillan, London 1972), p. 11.
21. F. R. Jevons, *The Teaching of Science* (Allen & Unwin, London 1969), p. 19.

Chapter 2
1. D. J. de Solla Price, preface to *Little Science, Big Science* (Columbia University Press 1965).
2. ibid., p. 7.
3. *Statistics of Science and Technology 1970*, table 1.
4. Rose and Rose, *Science and Society*, p. 107.
5. Price, *Little Science, Big Science*, pp. 36–8.
6. ibid., pp. 41–9.
7. E. Garfield, *Nature* vol. 227 (1970), p. 669.
8. A brief account is given in F. R. Jevons, *The Biochemical Approach to Life* (Allen & Unwin, London 1968), pp. 24–7.

Chapter 3
1. T. Roszak, *The Making of a Counter Culture* (Faber, London 1970), p. 8.
2. ibid., pp. 214–15.
3. ibid., p. 208.
4. ibid., pp. 52–3.
5. ibid., p. 69 and p. 206.
6. B. Crick, *In Defence of Politics* (Penguin Books, Harmondsworth 1964), p. 21.
7. Roger Williams, *Politics and Technology*, p. 27.
8. Commission on the Third London Airport, *Report* (HMSO, London 1971).
9. The easiest entry points to Popper's own writings are chapter 1 of *Conjectures and Refutations*, 3rd edn (Routledge & Kegan Paul, London 1969); and chapter 1 of *The Logic of Scientific Discovery* (Hutchinson, London 1959).
10. P. B. Medawar, *The Art of the Soluble* (Penguin Books, Harmondsworth 1969), p. 130. See also his *Induction and Intuition in Scientific Thought* (Methuen, London 1969), for a fuller treatment.
11. J. C. Eccles, quoted in Popper, *Conjectures and Refutations*, p. 2.
12. A more sophisticated treatment is given by I. Lakatos in I. Lakatos and A. Musgrave (eds), *Criticism and the Growth of Knowledge* (Cambridge University Press 1970), p. 91.
13. A. Koestler, *The Sleepwalkers* (Penguin Books, Harmondsworth 1964), pp. 195 and 579.
14. J. D. Watson, *The Double Helix* (Penguin Books, Harmondsworth 1970), p. 152.
15. Popper, *Conjectures and Refutations*, p. 38.
16. Watson, *The Double Helix*, p. 134.

Chapter 4
1. T. S. Kuhn, *The Copernican Revolution* (Harvard University Press 1957).

2. Lakatos and Musgrave (eds), *Criticism and the Growth of Knowledge.*
3. Popper in *Criticism and the Growth of Knowledge*, p. 51.
4. T. S. Kuhn, *The Structure of Scientific Revolutions*, 2nd edn (Chicago University Press 1970), p. 10. This edition is almost identical with the 1st edition of 1962 except for the addition of a postscript.
5. ibid., p. 11.
6. M. Masterman in *Criticism and the Growth of Knowledge*, p. 59.
7. Kuhn in *Criticism and the Growth of Knowledge*, p. 231, and especially p. 272.
8. Kuhn, *The Structure of Scientific Revolutions*, p. 12.
9. ibid., p. 13.
10. ibid., p. 20.
11. ibid., p. 37.
12. ibid., p. 164.
13. Medawar, *The Art of the Soluble*, p. 97.
14. ibid., p. 137.
15. Kuhn, *The Structure of Scientific Revolutions*, p. 65.
16. M. Polanyi, *Science, Faith and Society* (Chicago University Press 1964), p. 29.
17. Kuhn, *The Structure of Scientific Revolutions*, p. 77.
18. N. R. Hanson, *Patterns of Discovery: an Inquiry into the Conceptual Foundations of Science* (Cambridge University Press 1958), chap. 1.
19. Kuhn, *The Structure of Scientific Revolutions*, p. 117.
20. Medawar, *The Art of the Soluble*, p. 159.
21. Kuhn, *The Structure of Scientific Revolutions*, p. 47.
22. Jevons, *The Teaching of Science*, p. 146.
23. Kuhn, *The Structure of Scientific Revolutions*, p. 138.
24. ibid., p. 2.
25. ibid., p. 108.
26. ibid., p. 155.
27. ibid., p. 158.
28. ibid., p. 151.
29. ibid., p. 170.
30. Lakatos in *Criticism and the Growth of Knowledge*, p. 178.
31. Kuhn in *Criticism and the Growth of Knowledge*, p. 261.
32. For a more optimistic view, see Popper, *Conjectures and Refutations*, chap. 3.

Chapter 5
1. Useful introductions to this relatively new field are provided by M. J. Mulkay, *The Social Process of Innovation* (Macmillan, London 1972); and by a collection of readings edited by B. Barnes, *Sociology of Science* (Penguin Books, Harmondsworth 1972).
2. W. O. Hagstrom, *The Scientific Community* (Basic Books, New York 1965), p. 6.
3. N. W. Storer, *The Social System of Science* (Holt, Rinehart & Winston, New York 1966), chap. 3.
4. Crick, *In Defence of Politics*, p. 21.
5. Medawar, *The Art of the Soluble*, p. 96.

6. Koestler, *The Sleepwalkers*, p. 382.
7. Hagstrom, *The Scientific Community*, p. 14.
8. Watson, *The Double Helix*, p. 129.
9. Quoted by Kuhn, *The Structure of Scientific Revolutions*, p. 153.
10. Hagstrom, *The Scientific Community*, p. 37.
11. This and other dichotomies are discussed by S. Cotgrove and S. Box, *Science, Industry and Society* (Allen & Unwin, London 1970), p. 24.
12. Hagstrom, *The Scientific Community*, p. 35.
13. M. Gibbons, quoted in F. R. Jevons, 'Problems facing University Science', *Nature* vol. 229 (1971), p. 601.
14. *Statistics of Science and Technology 1970*, table 34.
15. Cotgrove and Box, *Science, Industry and Society*.
16. Jevons, *The Teaching of Science*, chap. 4.
17. D. S. Davies in F. R. Jevons and H. D. Turner (eds), *What Kinds of Graduates do we Need?* (Oxford University Press 1972), p. 40.
18. C. C. Butler in *What Kinds of Graduates do we Need?*, p. 81.
19. Jevons, *The Teaching of Science*, chap. 5; and in *What Kinds of Graduates do we Need?*, p. 95.
20. M. C. McCarthy, Annex F to the Swann Report. Committee on Manpower Resources for Science and Technology, *The Flow into Employment of Scientists, Engineers and Technologists* (Cmnd 3760, HMSO, London 1968).
21. *Statistics of Science and Technology 1970*, table 49.

Chapter 6
1. Reprinted in A. Johnston (ed.), *Francis Bacon* (Batsford, London 1965), pp. 13–15.
2. This summary is taken from Jevons, *The Teaching of Science*, pp. 67–8.
3. G. Basalla (ed.), *The Rise of Modern Science: Internal or External Factors?* (D. C. Heath, Lexington, Massachusetts 1968).
4. R. K. Merton, 'Science and Economy of Seventeenth-Century England', in B. Barber and W. Hirsch (eds), *The Sociology of Science* (Free Press, Glencoe, Illinois 1962), pp. 67–88.
5. This summary is taken from Jevons, *The Teaching of Science*, pp. 51–2.
6. F. Engels, quoted in Barnes, *Sociology of Science*, p. 17.
7. *Framework for Government Research and Development* (Cmnd 5046, HMSO, London 1972), para. 50.
8. ibid., para. 5.
9. ibid., paras 13–27.
10. S. C. Gilfillan, *The Sociology of Invention* (Follett Publishing Co., Chicago 1935), p. 7.
11. P. M. S. Blackett, *Nature* vol. 219 (1968), p. 1107.
12. J. H. Hollomon, in R. A. Tybout (ed.), *Economics of Research and Development* (Ohio State University Press, Columbus 1965), p. 253.
13. M. Gibbons and C. F. Johnson, *Nature* vol. 227 (1970), p. 125.
14. J. Langrish, *Science Journal* (Dec. 1969), p. 81.
15. B. R. Williams, *Technology, Investment and Growth* (Chapman & Hall, London 1967), chap. 4.
16. Science Policy Research Unit, University of Sussex, *Success and*

Failure in Industrial Innovation (Centre for the Study of Industrial Innovation, London 1972), p. 5.

Chapter 7

1. Gordon Rattray Taylor, *The Doomsday Book* (Thames & Hudson, London 1970); 'A Blueprint for Survival', *The Ecologist* vol. 2 no. 1 (Jan. 1972); R. J. Dubos and B. Ward, *Only One Earth: the Care and Maintenance of a Small Planet* (Deutsch, London 1972).
2. D. H. Meadows, D. L. Meadows, J. Randers and W. W. Behrens III, *The Limits to Growth* (Earth Island Ltd, London 1972), p. 124.
3. ibid., p. 127.
4. ibid., p. 163.
5. ibid., p. 145.
6. F. Felix, *World Markets of Tomorrow* (Harper & Row, London 1972).
7. Jones, *The Role of Science and Technology in Developing Countries*, p. 3. Based on World Bank data for 1967.
8. W. Beckerman, 'Economist, Scientists, and Environmental Catastrophe', *Oxford Economic Papers* vol. 24 (1972), p. 327.
9. J. Maddox, *The Doomsday Syndrome: an Assault on Pessimism* (Macmillan, London 1972).
10. *Pollution: Nuisance or Nemesis?*, paras 24–50, 227–40.
11. J. H. Barnett and C. Morse, *Scarcity and Growth: the Economics of Natural Resource Availability* (Johns Hopkins Press, Baltimore 1965), pp. 32 and 48.
12. Meadows *et al.*, *The Limits to Growth*, pp. 56–60.

APPENDIX 1

The Social Function of Science: After Twenty-five Years

J. D. BERNAL

From M. Goldsmith and A. Mackay (eds), *The Science of Science* (Penguin 1966), p. 285

Twenty-five years after writing *The Social Function of Science* it is interesting to look back to see how far its thesis was justified and how far any of its lessons have been learned and still hold any message for the present or the future. I would now conclude that to a very large extent the book has fulfilled its original object: to make people aware of the new function that science was acquiring then and would increasingly acquire in the future, in determining the conditions of human life and – as it is now tragically revealed – of the very existence of humanity. The events that followed very soon after its publication were to bring this home to everyone.

We are no longer concerned, as I was then, merely to vindicate the growth and use of science in modern civilisation. It is there for bad or good, hence it is even more essential to understand it. In *The Social Function*, that was what I was trying to do. Yet I failed to foresee how rapidly the tendencies I had observed were to bear fruit and to what extent the prophecy I made at that time was to be fulfilled and over-fulfilled. . . .

Now the 'research revolution', to borrow the title of Mr Silk's fascinating and horrifying study, is not only a fact but a recognised fact of the time. The economics of modern states are no longer considered to be economics of a fluctuating equilibrium but economics of growth. The rate of growth of the gross national product is now taken as the index of national economic health or even of survival among the advanced industrial countries. To achieve a merely tolerable rate of increase of the national product, say about 4 per cent, depends in the first place on the amount of past research that can be applied at the time; but also the rate of increase in the future depends on the amount of research that is carried out now. Furthermore, the time lag of the application of research has greatly shortened; new ideas can come into application, especially in fields which are advancing most rapidly, like those of control mechanism, within a year or two of their first discovery.

The recognition of this led, first in the military scientific field, to a research race which is still going on and has spread now into the civil field, not only in the electrical and chemical industries but also in biology, medicine, and agriculture. In the years since *The Social Function* was written the yield per man in agriculture has multiplied three times and, correspondingly, the number of persons directly involved in agriculture has shrunk – at present only $2\frac{1}{2}$ per cent of the population of the United States and only 5 per cent in Britain. This is at a time when more than 70 per cent are occupied in agriculture in the poorer parts of the world; the difference marks a real practical achievement of the scientific revolution.

This very success, however, also marks the failure of the research revolution to spread effectively over the two-thirds of the world which are just struggling out of the old colonial regimes. The gap between the economies of the advanced industrial countries and those of the developing countries is rapidly widening. Only a small part of this can be put down to the increase in population in the developing parts of the world. Even with the increase as it is running today, at about 2 per cent per annum, this is so much less than the rate of increase in the scientific potential, which comes to more like 20 per cent per annum, that there can be no question of an automatic and independent 'catch-up' on the part of the developing countries.

Whether the gap is filled or not, however, we cannot ignore the threat of utter destruction that one aspect of the scientific revolution holds over all mankind, the destruction typified by the fission and now the fusion bomb. Concern with war has dominated the gigantic scientific efforts of the last twenty years. It has unquestionably affected the new scientific revolution, which it first stimulated and then hampered with its demands on manpower and apparatus.

Enormous changes have occurred since *The Social Function* was written, comparatively only a few years ago. It was written on the eve of the Second World War, with its vast destruction and the liberations that it brought about, especially in Asia and Africa; but more significant than the constructive aspects was the discovery of nuclear fission culminating in the atom bomb and the threat which it implies to the whole of life. In mastering the atom something of the full power of science was made manifest, but what is equally obvious is that the powers which controlled humanity at that time, political and financial, were incapable of using the potentialities of science. They were incapable, really, even of understanding it, and the twenty years we have been spared in the atomic age are only now bringing the lesson home.

If we can survive the dangers of the immediate present we have every chance of realising a world so different from anything we have had before that the transition is greater than any which has occurred since the first appearance of humanity. We have the potentiality of the age of abundance and leisure, but the actuality of a divided world with greater poverty, stupidity, and cruelty than it has ever known.

Between that world and the present, however, we clearly have to pass through a transitional period which will be one of great danger. The technical possibilities and, even more, the integrated control that can be achieved through the proper use of computers, cannot be fitted into the fragmented social frame of private interests and exploitation. The operational problem remains of how to effect the transition with the minimum of strain and destruction. I feel confident that the ultimate pattern will, so to speak, impose itself the moment its logic is fully appreciated, but I do not minimise the danger of at least some parts of the new scientific method, especially those of mass communication and education, being used to retard this change or to deflect it.

I wrote *The Social Function* just before the Second World War. In this war the ideas that were exposed there and that were then largely theoretical were fully tested in practice. It was possible in the service of war to carry out many of the proposals for organised science and its application that I had made in the book. I summarised some of these in my paper on 'The Lessons of the War for Scientists' (see J. D. Bernal, *The Freedom of Necessity*, London 1949): 'The freedom of scope for experimentation and assistance is a lesson which will not be lost on the scientists who experienced it. It will be of particular importance in the next few years when we are bound to suffer for lack of men to carry out the very much increased tasks which will have to be dealt with by science and where it is more than ever important that we should make the fullest use of our few capable workers. The principle first enunciated by Professor Blackett, that allocation of money to science should be made in the measure of what a competent scientist can usefully spend and not according to what he can just manage on, should be the basis for our post-war science.

'Almost equally important as a lesson of the war is the value of the greater integration which was achieved in scientific work, partly through a more rational organisation and partly through the function of an effective positive information service.'

The organization of science in wartime 'provided what had been previously the function of the scientific societies, that is, careful discussion and interchange of scientific opinion, but it had also a much more positive function in laying out lines of attack and in determining priorities. In this way the scientific work itself could be carried out in a multiplicity of actual experimental stations – government, industrial, university – and yet not lose its coherence or general direction. Further, this direction was exercised by scientists themselves, at least in the latter part of the war, and was consequently sufficiently reasonable to be for the most part acceptable to the main body of scientific workers. Out of it emerged general concepts of organisation of science which will be of permanent value' (p. 290).

One major result of science in war was the foundation of Operational Research. 'Operational research [I wrote], led not only to greater understanding in detail of the operations of war, but to a much clearer integration of different types of operations. As the war went on, combined operations, whether by land and sea, land and air, or all three together, became the rule rather than the exception and the bridge between the very diverse approaches of the different services was often effected through operational research. In this way several general principles emerged which had far wider application than merely to military operations (p. 297).

'The original implications of operational research are already making themselves felt in peacetime economy. In principle it amounts to the statement that any human activity and any branch of that activity is a legitimate subject for scientific study, and subsequently for modification in the light of that study. Once this is accepted in practice, which implies the provision of research workers to carry out these studies, the way is open to a new level of man's control of his environment, one in which economic and social processes become scientific through and through. This is already happening in productive industry. We are witnessing what is really a new industrial revolution in which statistical and scientific control and rational planning and design are taking the place that prime movers and simple mechanism did in the first industrial revolution. Industrial processes are now seen to represent cycles of performance in which the needs of the consumer determine production and are in turn modified by the results of that production, leading to a progressively greater degree of satisfaction at a steadily diminishing social cost' (p. 299).

The direction implied in my major conclusion still holds. 'The most balanced and flexible plan for scientific research, however, will not be enough. It will need to be integrated with a positive drive: a technical, biological and social advance carried out with all the resources of the community. That such a task can be achieved has been shown by the experience of the war; but the war has shown also that it is not only possible but absolutely necessary for survival as an advanced community. A national economy, integrated through science and continually advancing by means of scientific research and development, is the basic need of the new era which we are now entering. It implies the expenditure of a much larger proportion of social effort and social resources on science than ever envisaged before.

Those who had considered the advantages that science could bring to society had realised well before the war that the expenditure on science by society was far too small; at that time the total expenditure in this country was something of the order of one-tenth of 1 per cent of the national income. They could see, and they tried to point out how the increase of this proportion would bring far more rapid prosperity. In the post-war situation, however, with the leeway of destruction and disorganisation to make up and the far weaker and even perilous situation of this country, what was desirable has become an absolute necessity, and the proportion to be aimed at must be a much higher one. . . . On the long-term view we must look forward to a fairly rapid transformation in which scientific functions – not necessarily scientific research and development only, but scientific production and scientific administration – will absorb a progressively larger and larger proportion of the population. From one-tenth of 1 per cent we may advance to involving 1, 2 and possibly ultimately, but in the far distant future, as much as 20 per cent of the population in such activities. This is a logical consequence of the increasing role of human intelligence and consciousness in the management of our society. Long before such a stage is reached, however, the distinction between scientific and non-scientific activity will probably have largely disappeared. Already we require for the proper functioning of our society a certain degree of knowledge of the facts of science and even more of its method on the part of every citizen. The government cannot make decisions, the people cannot carry out the decisions reached, unless they have much fuller understanding than at present of what they are doing' (pp. 308–9). . . .

The development of the computer . . . is an example of the disproportion between the essential scientific feature of an invention and its utility. The mathematical notions behind the modern computer are no more complex than those of the computer first designed by Pascal in the seventeenth century and partly executed by Babbage in the nineteenth. What brought the idea to life again was the means to carry it out: the components, no longer wooden cogs or even metal ones, as in the earlier machines, were electrical circuits very rapidly switched, first by means of valves and magnetic circuits, and finally by means of semiconductors. The result was not an invention of any one person – it did not require genius, but simply application of known methods to known problems. But once it came, it was to have enormous effects which are only just beginning to be seen. . . .

All these great achievements – in power, in industry, in medicine and agriculture – are themselves only a part of what is now being seen more consciously to be the major transformation of our time, the research revolution itself. We have now arrived at the second stage, that of the development of the scientific method. To quote Bacon: 'But above all, if a man could succeed, not in striking out some particular invention, however useful, but in kindling a light in Nature – a light which should in its very rising touch and illuminate all the border-regions that confine upon the circle of our present knowledge; and so spreading further and further should presently disclose and bring into sight all that is most hidden and secret in the world – that man (I thought) would be the benefactor indeed of the human race – the propagator of man's empire over the universe, the champion of liberty, the conqueror and subduer of necessities.' Bacon was talking about the scientific method itself. What has happened recently is the realisation not only by scientists, who have known it for many years, but also by peoples and governments, that here is a method which in itself can be *counted on* to generate more and more of these great achievements and transformations. This is the deeper meaning of the research revolution. That revolution has begun, and it is going on faster and faster.

But that is only half the story. Research can be carried out and applied in a most disorderly and wasteful way. In *The Social Function* I estimated the efficiency of research as about 2 per cent, that is, about 2 per cent of what could have been found out and been done with the resources and men available was in fact carried out. To achieve even modest increases in efficiency, obviously something else, but something radically different is needed. We need a strategy for research which must be based on a *science of science*. This cannot be formulated by merely laying down *a priori* what the scientific method should be, as in the past, but by finding this out from what it does, through its modes of action. These modes now involve machines as well as human beings. The science of science, or the self-consciousness of science, as I have put it elsewhere, is the real drastic advance of the second part of the twentieth century. This science of science must be wide-ranging; it must include the social and economic as well as the material and technical conditions for scientific advance and for the proper use of its tools. . . .

Here again, the strictures I put on existing systems of scientific education, much resented as they were at the time, now appear almost as commonplace in the light of the new and urgent requirements for scientific and technological manpower that are put forward not only in the industrial states but in the developing ones as well.

There is no denying the existence of the problem. It has . . . already led to a basic change in emphasis in education, away from the

Renaissance ideal of producing a cultivated elite, to one aiming at producing administrators and possibly even governors capable of understanding and appreciating the needs of science in an industrial developing society. But the problem is far from being solved. Indeed, on the face of it it would appear to be insoluble. The amount of time available for education can be stretched only to a very limited extent, doubling it, for instance, from three years to six, but, with the output of science doubling every seven years, it is clear that entirely new methods of teaching will have to be evolved to make use of the knowledge already acquired and even more to ensure the continued rapidity of the acquisition and integration of new knowledge.

However, here the new techniques of the computer age can help. Already teaching machines are being evolved that can adapt themselves to the speed of learning of individual students, and techniques of television can also supplement to a large extent practical instruction. Here again, however, nothing effective will be done unless very much effort is devoted to research on methods of teaching science. There is a realisation, which is just beginning in some of the older industrial countries that not only a small section of a professional class needs such education, but that it must also be spread throughout the whole population. Modern automatic machinery requires highly educated personnel to watch its working, and to deduce from its performance the best way of improving it. In any case it is clear that the requirements for personnel in research and development in industry, agriculture, and medicine will be enormously increased and come to equal and in some cases surpass the number of people involved in the operation of machinery and transport. Thus the development of automation, far from decreasing the need for science, will actually increase it many fold. . . .

The problem of transforming the world to take account of the scientific revolution is everywhere a difficult one, and one which for the moment is growing more difficult with time; but this can only be a temporary phase. The whole problem – economic, scientific, and political – must be regarded as one of a planned operation, definite phases of advance to be kept in step by some kind of international co-ordination. Whether such a co-ordination is possible in a world dominated by divisions between capitalist and socialist economies is the great problem of our time. If the negative view is taken, as it is in China, it might appear that two radically different kinds of science will grow up in parallel, one gradually dominating and the other shrinking away. If, on the other hand, the possibility of coexistence and, even more, co-operation is admitted it may be possible to move step by step from the very limited international co-operation in science that exists today to a more complete one; it will be all the easier when the levels of production and technical advance and the political and economic systems come nearer together. Only time can resolve this difficulty, but the scientists of the world today must act on the best analysis that they can now make and push as hard as they can for the most international organisation of science that can be achieved. . . .

The Social Function was permeated with the picture of the frustration of science arising, in the first place, largely from the financial stringency in which it worked. Much of my book was taken up with arguing against this limitation. Now there is a different situation – it is the large scale of expenditure on science rather than a small scale that must be considered. During the period of the war and for the first ten years afterwards much of the expenditure on science, being predominantly military in character, was subject to very convenient principles of military finance – all sums asked for are granted, and if any questions are raised the questioner is told that for reasons of security no further information can be given. How the money was allocated and to whom were matters of state secrets. Parliaments were expected to pass military budgets and to raise the new taxation without question. It was considered that military science was sacrosanct.

This situation no longer quite exists. The immediate danger seems to have receded, though in fact the military budget is continually increasing. Now, however, some daring legislators, even in the United States are beginning to question what happens to the money. There is a definite tendency to cut scientific expenditure or, at least, to hold back its unlimited growth. . . .

At the present moment I feel that we are grossly underplaying the use of fundamental science. The quickest and also the surest returns would come from a deeper understanding of nature. Much of so-called applied science is applied obsolete science; the methods of application are even more obsolete than the science they apply. For example, building is admittedly one of the most backward parts of modern technique. Because we do not know enough about either the strength of the materials that we are using or about the reactions to the stresses, which have hitherto been incalculable, we put anything like ten times more material into the production of useful space than we need. It is called a factor of safety: it is really one of ignorance. Additional knowledge would pay enormous dividends, and yet the amount spent on fundamental research in this field is practically negligible. Of course, there are various reasons for this. The weight of technological education, the concept of good practice, together with the idea that the profits of the building industry depend on how much material is used and how slowly the building can be put up, all stand in the way. We are still using the bricks that were good enough for our Babylonian predecessors, each laboriously laid by hand. Building needs to be mechanised before it can be automised and brought into harmony with modern industry. The technical advances which I anticipate will inevitably imply economic changes of a basic nature. The scientific and computer age is necessarily a Socialist one. . . .

For scientists there has grown up, especially since the last war, a number of new initiatives, based less on the idea of the position of the scientist in production than on the responsibility of the scientist for the military developments of our time, particularly for the horror of the atomic and hydrogen bombs. This has given rise to a much greater consciousness of scientists, exemplified by the movements of the Pugwash Committee, arising out of the Einstein–Russell letter in July 1955, and by the parallel movements such as that initiated by Linus Pauling, the Society for the Social Responsibility of Scientists. There is no doubt that, although the membership of these groups is at the moment limited, their views are much more widely shared, and it is only fear or caution that prevents the great majority of scientists from expressing them. The important thing in this is not so much the attitude of individual scientists as the collective effort to block out at least ideal policies which would have the general direction of making science serve the preservation and not the destruction of humanity. The more scientific effort that is directed to military ends, the more resistance it will create in the minds of scientists. The awareness of the proper use of science in society is not easy to reach, and it is harder still to get agreement on it even among scientists. The scientist as citizen is not in the first place a scientist, only in the second. In the course of discussions in these and other movements he becomes aware that it is necessary to have a unitary outlook, that he cannot be torn apart by the contradictions between his science and his duty. He sees a world in which the use of science has become the dominating factor. Mankind cannot progress, cannot even exist today without science. However, far from giving him a sense of power, it emphasises his awareness of his present weakness and futility. The powers of ignorance and greed distort science and lead it astray for war and destructive ends. . . .

Criteria for Scientific Choice

A. M. WEINBERG

From *Minerva* vol. 1 (1963), p. 159

I believe that criteria for scientific choice can be identified. In fact, several such criteria already exist; the main task is to make them more explicit. The criteria can be divided into two kinds: internal criteria and external criteria. Internal criteria are generated within the scientific field itself and answer the question: How well is the science done? External criteria are generated outside the scientific field and answer the question: Why pursue this particular science? Though both are important, I think the external criteria are the more important.

Two internal criteria can be easily identified: (i) Is the field ready for exploitation? (ii) Are the scientists in the field really competent? Both these questions are answerable only by experts who know the field in question intimately, and who know the people personally. These criteria are therefore the ones most often applied when a panel decides on a research grant: in fact, the primary question in deciding whether to provide governmental support for a scientist is usually: How good is he?

I believe, however, that it is not tenable to base our judgements entirely on internal criteria. As I have said, we scientists like to believe that the pursuit of science as such is society's highest good, but this view cannot be taken for granted. For example, we now suffer a serious shortage of medical practitoners, probably to some extent because many bright young men who would formerly have gone into medical practice now go into biological research: government support is generally available for post-graduate study leading to the Ph.D. but not for study leading to the medical degree. It is by no means self-evident that society gains from more biological research and less medical practice. Society does not *a priori* owe the scientist, even the good scientist, support any more than it owes the artist or the writer or the musician support. Science must seek its support from society on grounds other than that the science is carried out competently and that it is ready for exploitation; scientists cannot expect society to support science because scientists find it an enchanting diversion. Thus, in seeking justification for the support of science, we are led inevitably to consider external criteria for the validity of science – criteria external to science, or to a given field of science.

Three external criteria can be recognised: technological merit, scientific merit and social merit. The first is fairly obvious: once we have decided, one way or another, that a certain technological end is worth while, we must support the scientific research necessary to achieve that end. Thus, if we have set out to learn how to make breeder reactors, we must first measure painstakingly the neutron yields of the fissile isotopes as a function of energy of the bombarding neutron. As in all such questions of choice, it is not always so easy to decide the technological relevance of a piece of basic research. The technological usefulness of the laser came after, not before, the principle of optical amplification was discovered. But it is my belief that such technological bolts from the scientific blue are the exception, not the rule. Most programmatic basic research can be related fairly directly to a technological end at least crudely if not in detail. The broader question as to whether the technological aim itself is worth while must be answered again partly from within technology through answering such questions as: Is the technology ripe for exploitation? Are the people any good? Partly from outside technology by answering the question: Are the social goals attained, if the technology succeeds, themselves worth while? Many times these questions are difficult to answer, and sometimes they are answered incorrectly: for example, the United States launched an effort to control thermonuclear energy in 1952 on a rather large scale because it was thought at the time that controlled fusion was much closer at hand than it turned out to be. Nevertheless, despite the fact that we make mistakes, technological aims are customarily scrutinised much more closely than are scientific aims; at least we have more practice discussing technological merit than we do scientific merit.

The criteria of scientific merit and social merit are much more difficult: scientific merit because we have given little thought to defining scientific merit in the broadest sense, social merit because it is difficult to define the values of our society. As I have already suggested, the answer to the question: Does this broad field of research have scientific merit? cannot be answered within the field. The idea that the scientific merit of a field can be judged better from the vantage point of the scientific fields in which it is embedded than from the point of view of the field itself is implicit in the following quotation from the late John von Neumann: 'As a mathematical discipline travels far from its empirical source, or still more, if it is a second and third generation only indirectly inspired by ideas coming from reality, it is beset with very grave dangers. It becomes more and more pure aestheticising, more and more purely *l'art pour l'art*. This need not be bad if the field is surrounded by correlated subjects which still have closer empirical connections or if the discipline is under the influence of men with an exceptionally well-developed taste. But there is a grave danger that the subject will develop along the line of least resistance, that the stream, so far from its source, will separate into a multitude of insignificant branches, and that the discipline will become a disorganised mass of details and complexities. In other words, at a great distance from its empirical source, or after much "abstract" inbreeding, a mathematical subject is in danger of degeneration. At the inception the style is usually classical; when it shows signs of becoming baroque, then the danger signal is up.'

I believe there are any number of examples to show that von Neumann's observation about mathematics can be extended to the empirical sciences. *Empirical* basic sciences which move too far from the neighbouring sciences in which they are embedded tend to become 'baroque'. Relevance to neighbouring fields of science is, therefore, a valid measure of the scientific merit of a field of basic science. In so far as our aim is to increase our grasp and understanding of the universe, we must recognise that some areas of basic science do more to round out the whole picture than do others. A field in which lack of knowledge is a bottleneck to the understanding of other fields deserves more support than a field which is isolated from other fields. This is only another way of saying that, ideally, science is a unified structure and that scientists, in adding to the structure, ought always to strengthen its unity. Thus, the original motivation for much of high energy physics is to be sought in its elucidation of low energy physics, or the strongest and most exciting motivation for measuring the neutron capture cross sections of the elements lies in the elucidation of the cosmic origin of the elements. Moreover, the discoveries which are acknowledged to be the most important scientifically have the quality of bearing strongly on the scientific disciplines around them. For example, the discovery of X-rays was important partly because it extended the electromagnetic spectrum but, much more, because it enabled us to see so much that we had been unable to see. The word 'fundamental' in basic science, which is often used as a synonym for 'important', can be partly paraphrased into 'relevance to neighbouring areas of science'. I would therefore sharpen the criterion of scientific merit by proposing that, other things being equal, *that field has the most scientific merit which contributes most heavily to and illuminates most brightly its neighbouring scientific disciplines*. This is the justification for my previous suggestion about making it socially acceptable for people in *related* fields to offer opinions on the scientific merit of work in a given field. In a sense, what I am trying to do is to extend to basic research a practice that is customary in applied science: a project director trying to get a reactor built on time is expected to judge the usefulness of component development and fundamental research which bears on his problems. He is not always right; but his opinions are usually useful both to the researcher and to the management disbursing the money.

I turn now to the most controversial criterion of all – social merit or relevance to human welfare and the values of man. Two difficulties face us when we try to clarify the criterion of social merit: first, who is to define the values of man, or even the values of our own society; and second, just as we shall have difficulty deciding whether a proposed research helps other branches of science or technology, so we will have even greater trouble deciding whether a given scientific or technical enterprise indeed furthers our pursuit of social values, even when those values have been identified. With some values we have little trouble: adequate defence, or more food, or less sickness, for example, are rather uncontroversial. Moreover, since such values themselves are relatively easy to describe, we can often guess whether a scientific activity is likely to be relevant, if not actually helpful, in achieving the goal. On the other hand, some social values are much harder to define: perhaps the most difficult is national prestige. How do we measure national prestige? What is meant when we say that a man on the moon enhances our national prestige? Does it enhance our prestige more than, say, discovering a polio vaccine or winning more Nobel Prizes than any other country? Whether or not a given achievement confers prestige probably depends as much on the publicity that accompanies the achievement as it does on its intrinsic value.

Among the most attractive social values that science can help to achieve is international understanding and cooperation. It is a commonplace that the standards and loyalties of science are transnational. A new element has recently been injected by the advent of scientific research of such costliness that now it is prudent as well as efficient to participate in some form of international cooperation. The very big accelerators are so expensive that international laboratories such as CERN at Geneva are set up to enable several countries to share costs that are too heavy for them to bear separately. Even if we were not committed to improving international relations we would be impelled to co-operate merely to save money.

Bigness is an advantage rather than a disadvantage if science is to be used as an instrument of international co-operation: a $500,000,000 co-operative scientific venture – such as the proposed 1,000 Bev intercontinental accelerator – is likely to have more impact than a $500,000 Van de Graaff machine. The most expensive of all scientific or quasi-scientific enterprises – the exploration of space – is, from this viewpoint, the best-suited instrument for international co-operation. . . .

APPENDIX 3

Two Conceptions of Science

P. B. MEDAWAR

From *The Art of the Soluble* (Penguin 1969), p. 130

Let me turn now to two serious but completely different conceptions of science, embodying two different valuations of scientific life and of the purpose of scientific inquiry. For dialectical reasons I have exaggerated the differences between them, and I do not suggest that anybody cleaves wholly to the one conception or wholly to the other.

According to the first conception, Science is above all else an imaginative and exploratory activity, and the scientist is a man taking part in a great intellectual adventure. Intuition is the main-spring of every advancement of learning, and *having ideas* is the scientist's highest accomplishment; the working out of ideas is an important and exacting but yet a lesser occupation. Pure science requires no justification outside itself, and its usefulness has no bearing on its valuation. 'The first man of science', said Coleridge, 'was he who looked into a thing, not to learn whether it could furnish him with food, or shelter, or weapons, or tools, or orna-ments, or *play-withs*, but who sought to know it for the gratification of knowing. . . .'

Science and poetry in its widest sense are cognate, as Shelley so rightly said. So conceived, science can flourish only in an atmosphere of complete freedom, protected from the nagging importunities of need and use, because the scientist must travel where his imagination leads him. Even if a man should spend five years getting nowhere, that might represent an honourable and perhaps even a noble endeavour. The Patrons of science – today the Research Councils and the great Foundations – should support men, not projects, and individual men rather than teams, for the history of science is for the most part a history of men of genius.

The alternative conception runs something like this: Science is above all else a critical and analytical activity; the scientist is pre-eminently a man who requires evidence before he delivers an opinion, and when it comes to evidence he is hard to please. Imagi-nation is a catalyst merely: it can speed thought but cannot start it or give it direction; and imagination must at all times be under the censorship of a dispassionate and sceptical habit of thought. Science and poetry are antithetical, as Shelley so rightly said.[1] Scientific research is intended to enlarge human understanding, and its usefulness is the only objective measure of the degree to which it does so; as to freedom in science, two world wars have shown us how very well science can flourish under the pressures of necessity. Patrons of science who really know their business will support projects, not people, and most of these projects will be carried out by teams rather than by individuals, because modern science calls for a consortium of the talents and the day of the individual is almost done. If any scientist should spend five years getting nowhere, his ambitions should be turned in some other direction without delay.

I have made the one conception a little more romantic than it really is, and the other a little more worldly, and to restore the balance I want to express the distinction in a different and, I think, more fundamental way.

In the romantic conception, truth takes shape in the mind of the observer: it is his imaginative grasp of *what might be true* that provides the incentive for finding out, so far as he can, what *is* true. Every advance in science is therefore the outcome of a speculative adventure, an excursion into the unknown. According to the opposite view, truth resides in nature and is to be got at only through the evidence of the senses: apprehension leads by a direct pathway to comprehension, and the scientist's task is essentially one of *discernment*. This act of discernment can be carried out according to a Method which, though imagination can help it, does not depend on the imagination: the Scientific Method will see him through.[2]

Inasmuch as these two sets of opinions contradict each other flatly in every particular, it seems hardly possible that they should both be true; but anyone who has actually done or reflected deeply upon scientific research knows that there is in fact a great deal of truth in both of them. For a scientist must indeed be freely imaginative and yet sceptical, creative and yet a critic. There is a sense in which he must be free, but another in which his thought must be very precisely regimented; there is poetry in science, but also a lot of book-keeping.

There is no paradox here: it just so happens that what are usually thought of as two alternative and indeed competing accounts of *one* process of thought are in fact accounts of the *two* successive and complementary episodes of thought that occur in every advance of scientific understanding. Unfortunately, we in England have been brought up to believe that scientific discovery turns upon the use of a method analogous to, and of the same logical stature as deduction, namely the method of *Induction* – a logically mechanised process of thought which, starting from simple declarations of fact arising out of the evidence of the senses, can lead us with certainty to the truth of general laws. This would be an intellectually disabling belief if anyone actually believed it, and it is one for which John Stuart Mill's methodology of science must take most of the blame. The chief weakness of Millian induction was its failure to distinguish between the acts of mind involved in discovery and in proof. It was an understandable mistake, because in the process of deduction, the paradigm of all exact and conclusive reasoning, discovery and proof may depend on the same act of mind: starting from true premises, we can derive and so 'discover' a theorem by reasoning which (if it has been carried out according to the rules) itself shows that the theorem must be true. Mill thought that his process of 'induction' could fulfil the same two functions; but, alas, mistakenly, for it is not the origin but only the *acceptance* of hypotheses that depends upon the authority of logic.

If we abandon the idea of induction and draw a clear distinction between *having an idea* and *testing it* or *trying it out* – it is as simple as that, though it can be put more grandly – then the antitheses I have been discussing fade away. Obviously 'having an idea' or framing a hypothesis is an imaginative exploit of some kind, the work of a single mind; obviously 'trying it out' must be a ruthlessly critical process to which many skills and many hands may contribute. The form taken by scientific criticism is obvious too: experimentation *is* criticism; that is, experimentation in the modern sense, according to which an experiment is an act performed to test a hypothesis, not in the old Baconian sense, in which an experiment was a contrived experience intended to enlarge our knowledge of what actually went on in nature. Bacon exhorted us, rightly too, not to speculate upon but actually to experiment with loadstone and burning glass and rubbed amber: *his* experiments answer the question 'I wonder what would happen if . . .?' Baconian experimentation is not a critical activity but a kind of creative play.

The distinction between – and the formal separateness of – the creative and the critical components of scientific thinking is shown up by logical dissection, but it is far from obvious in practice because the two work in a rapid reciprocation of guesswork and checkwork, proposal and disposal, *Conjecture and Refutation*. Though imaginative thought and criticism are equally necessary to a scientist, they are often very unequally developed in any one man. Professional judgement frowns upon extremes. The scientist who devotes his time to showing up the inadequacies of the work of others is suspected of lacking ideas of his own, and everyone soon loses patience with the man who bubbles over with ideas which he loses interest in and fails to follow up.

The general conception of science which reconciles, indeed literally joins together, the two sets of contradictory opinions I have just outlined is sometimes called the 'hypothetico-deductive' conception. For our present clear understanding of the logical structure and wider scientific implications of the hypothetico-deductive system we are of course indebted to Karl Popper's *Logik der Forschung* of 1934, translated into English as *The Logic of Scientific Discovery*.

Everything I have said so far about the hypothetico-deductive system applies with exactly the same force to 'applied' science, even in its simplest and most familiar forms, as to that which is commonly called 'pure' or 'basic'. Imaginative conjecture and criticism, in that order, underlie the physician's diagnosis of his patient's ailments or the mechanic's explanation of why a car won't run. The physician may like to think himself, as Darwin did, an inductivist and a good Baconian, but with equally little reason, for Darwin was no inductivist; no more is he. . . .

Notes
1. This is not a debating point, for Shelley's writing can sustain both views. In his *Defence of Poetry* Shelley defines poetry in a 'universal' sense that comprehends all forms of order and beauty, and includes, therefore, not merely poetry in the narrower sense, but science as well (poetry 'comprehends all science'). Earlier, however, Shelley put Reason and Imagination at opposite poles; if then, as in the second conception I outline, science is regarded as an essentially rational activity, Shelley may quite rightly be allowed to speak for the view that science and poetry are antithetical.
2. For the conception that *truth is manifest*, see the critical analysis by K. R. Popper, 'On the Sources of Knowledge and of Ignorance', in *Conjectures and Refutations.* . . .

APPENDIX 4

Logic of Discovery or Psychology of Research?

T. S. KUHN

From I. Lakatos and A. Musgrave (eds), *Criticism and the Growth of Knowledge* (Cambridge University Press 1970), p. 4

Among the most fundamental issues on which Sir Karl [Popper] and I agree is our insistence that an analysis of the development of scientific knowledge must take account of the way science has actually been practised. That being so, a few of his recurrent generalisations startle me. One of these provides the opening sentences of the first chapter of the *Logic of Scientific Discovery*: 'A scientist,' writes Sir Karl, 'whether theorist or experimenter, puts forward statements, or systems of statements, and tests them step by step. In the field of the empirical sciences, more particularly, he constructs hypotheses, or systems of theories, and tests them against experience by observation and experiment.' The statement is virtually a cliché, yet in application it presents three problems. It is ambiguous in its failure to specify which of two sorts of 'statements' or 'theories' are being tested. That ambiguity can, it is true, be eliminated by reference to other passages in Sir Karl's writings, but the generalisation that results is historically mistaken. Furthermore, the mistake proves important, for the unambiguous form of the description misses just that characteristic of scientific practice which most nearly distinguishes the sciences from other creative pursuits.

There is one sort of 'statement' or 'hypothesis' that scientists do repeatedly subject to systematic test. I have in mind statements of an individual's best guesses about the proper way to connect his own research problem with the corpus of accepted scientific knowledge. He may, for example, conjecture that a given chemical unknown contains the salt of a rare earth, that the obesity of his experimental rats is due to a specified component in their diet, or that a newly discovered spectral pattern is to be understood as an effect of nuclear spin. In each case, the next steps in his research are intended to try out or test the conjecture or hypothesis. If it passes enough or stringent enough tests, the scientist has made a discovery or has at least resolved the puzzle he had been set. If not, he must either abandon the puzzle entirely or attempt to solve it with the aid of some other hypothesis. Many research problems, though by no means all, take this form. Tests of this sort are a standard component of what I have elsewhere labelled 'normal science' or 'normal research', an enterprise which accounts for the overwhelming majority of the work done in basic science. In no usual sense, however, are such tests directed to current theory. On the contrary, when engaged with a normal research problem, the scientist must *premise* current theory as the rules of his game. His object is to solve a puzzle, preferably one at which others have failed, and current theory is required to define that puzzle and to guarantee that, given sufficient brilliance, it can be solved. Of course the practitioner of such an enterprise must often test the conjectural puzzle solution that his ingenuity suggests. But only his personal conjecture is tested. If it fails the test, only his own ability not the corpus of current science is impugned. In short, though tests occur frequently in normal science, these tests are of a peculiar sort, for in the final analysis it is the individual scientist rather than current theory which is tested.

This is not, however, the sort of test Sir Karl has in mind. He is above all concerned with the procedures through which science grows, and he is convinced that 'growth' occurs not primarily by accretion but by the revolutionary overthrow of an accepted theory and its replacement by a better one. (The subsumption under 'growth' of 'repeated overthrow' is itself a linguistic oddity whose *raison d'être* may become more visible as we proceed.) Taking this view, the tests which Sir Karl emphasises are those which were performed to explore the limitations of accepted theory or to subject a current theory to maximum strain. Among his favourite examples, all of them startling and destructive in their outcome, are Lavoisier's experiments on calcination, the eclipse expedition of 1919, and the recent experiments on parity conservation. All, of course, are classic tests, but in using them to characterise scientific activity Sir Karl misses something terribly important about them. Episodes like these are very rare in the development of science. When they occur, they are generally called forth either by a prior crisis in the relevant field (Lavoisier's experiments or Lee and Yang's) or by the existence of a theory which competes with the existing canons of research (Einstein's general relativity). These are, however, aspects of or occasions for what I have elsewhere called 'extraordinary research', an enterprise in which scientists do display very many of the characteristics Sir Karl emphasises, but one which, at least in the past, has arisen only intermittently and under quite special circumstances in any scientific speciality.

I suggest then that Sir Karl has characterised the entire scientific enterprise in terms that apply only to its occasional revolutionary parts. His emphasis is natural and common: the exploits of a Copernicus or Einstein make better reading than those of a Brahe or Lorentz;

Sir Karl would not be the first if he mistook what I call normal science for an intrinsically uninteresting enterprise. Nevertheless, neither science nor the development of knowledge is likely to be understood if research is viewed exclusively through the revolutions it occasionally produces. For example, though testing of basic commitments occurs only in extraordinary science, it is normal science that discloses both the points to test and the manner of testing. Or again, it is for the normal, not the extraordinary practice of science that professionals are trained; if they are nevertheless eminently successful in displacing and replacing the theories on which normal practice depends, that is an oddity which must be explained. Finally, and this is for now my main point, a careful look at the scientific enterprise suggests that it is normal science, in which Sir Karl's sort of testing does not occur, rather than extraordinary science which most nearly distinguishes science from other enterprises. If a demarcation criterion exists (we must not, I think, seek a sharp or decisive one), it may lie just in that part of science which Sir Karl ignores.

APPENDIX 5

Social Control in Science

W. O. HAGSTROM

From *The Scientific Community* (Basic Books 1965), pp. 52–5 and 152–3

The thesis presented here is that social control in science is exercised in an exchange system, a system wherein gifts of information are exchanged for recognition from scientific colleagues. Because scientists desire recognition, they conform to the goals and norms of the scientific community. Such control reinforces and complements the socialisation process in science. It is partly dependent on the socialisation of persons to become sensitive to the responses of their colleagues. By rewarding conformity, this exchange system reinforces commitment to the higher goals and norms of the scientific community, and it induces flexibility with regard to specific goals and norms. The very denial by scientists of the importance of recognition as an incentive can be seen to involve commitments to higher norms, including an orientation to a scientific community extending beyond any particular collection of contemporaries.

At the inception of this research, it was thought possible to test the exchange theory almost directly. As has been indicated, scientists can more or less order the problems open to them by their importance in relation to other outstanding problems, and they can similarly order techniques by the degree to which they approach ideals of precision, and so forth. This rank-ordering of problems and techniques will be more or less common to members of a discipline, and scientists oriented to recognition from their colleagues may tend to select important problems and valued techniques. As a student said: 'Mathematicians want their work appreciated by others, especially by other mathematicians. You don't generally do things which won't be appreciated. One rule for usefulness is to get results which are useful in terms of other mathematical work.'

If one obtained a list of alternative problems in a discipline and their relative importance in the eyes of scientists in that discipline, one might be able to test such hypotheses as these: (i) scientists with the highest prestige tend to have solved problems in areas considered important by their colleagues; and (ii) when scientists shift from one problem area to another within a single discipline, they tend to shift from areas of low prestige to areas of higher prestige. If these two hypotheses were confirmed, the exchange theory would receive almost direct support.

Unfortunately, it proved impossible to obtain the data with which to test these hypotheses. In the first place, scientists were reluctant to discuss the importance attached by their colleagues to their problems. For example, a physical chemist was asked, '[Is your] research topic of central or of peripheral interest among physical chemists?' He replied, 'I have never given it a thought.' Another physical chemist replied to a subtler version of this question ('Do you think more physical chemists should be interested in your problem?') with 'I couldn't say.' Such responses are to be expected if the prestige of his specialty is important to a scientist, for it would amount to asking him about his own prestige, and for the most part people resist answering questions of this kind. Of course, it was often possible to determine the relative perceived importance of different fields. For example, solid-state physicists generally agreed that solid-state physics is considered less important than nuclear physics, and mathematicians generally agreed that algebraic topology has high prestige among American mathematicians. But even here perceptions were often biased by the context in which the informant worked. For example, mathematicians working in departments where analysis is given much importance were more likely than others to mention functional analysis as a field of central interest to mathematicians generally.

In the same way, it is difficult to obtain good ratings of the prestige of individuals from a small sample of scientists in a few departments. However, if larger samples of respondents were presented with fixed lists of fields and individuals, one might obtain the data with which to test these two hypotheses. This would provide further confirmation of the information-recognition exchange theory of scientific organisation.

At present, most of the evidence in support of the theory presented here is necessarily indirect. Such indirect support may consist of the rejection of competing theories or the demonstration of propositions more or less implied by the theory. The primary points of view – I shall call them 'theories' although they have not been sufficiently elaborated to justify the term that compete with the information-exchange theory stress either socialisation or the importance of material rewards. The naive individualism that denies the necessity for any process of social control has already been criticised. The other competing theory is more or less its opposite. Whereas naive individualism holds that scientists are not motivated by a desire for extrinsic rewards of any kind, the other theory is that scientists are motivated by the same kinds of extrinsic rewards as 'everybody else', namely, position and money. This would assert that the decisions of scientists are determined by the authorities who control these rewards – that scientists publish and work on certain topics and with certain techniques rather than others because only if they do so will they be rewarded by the higher authorities. This could be called a 'contractual' theory of the organisation of science.

Science is, in fact, an occupation and scientists do receive material rewards, and these are undoubtedly important. The problem for the information-recognition exchange theory is to determine if recognition and material rewards are actually consistent, and, if so, why. In this chapter, evidence has been provided in support of the view that the two types of rewards tend to be consistent and that material rewards tend to reinforce the operation of a system in which information is exchanged for recognition. The contractual theory contradicts this in two not entirely consistent ways. First, it asserts that material rewards are more important than recognition, that, for example, recognition is more likely to follow appointment to a distinguished faculty than to precede it. Second, it implies that this is the only way in which a system could be organised – although this implication is seldom made explicit. The theory appears to be mistaken in both respects. The most obvious evidence is the fact that many scientists who are under no coercive pressures to publish do so anyway. This is true, for example, of professors with tenure in large universities and of scientists in industrial research laboratories. Furthermore, as has been noted, organisational pressures on scientists to publish are more likely to reinforce the power of the scientific community than to supplant it. There is little evidence that university officials have much influence over the choices of problems and methods of university scientists. The power of the colleague community is also reinforced by the elaborate role-set possessed by most scientists.

The contractual theory of scientific organisation is an incomplete theory. It may account for publications, but it does not account for the selection of problems or methods. It might be argued that agencies which allocate research funds do this. However, evidence has been presented that scientists often control grant-giving agencies, so that the agencies tend to use the same criteria for evaluating scientists as do scientific colleagues when they award recognition. The contractual theory, unable to explain how 'good' problems are selected and 'bad' ones avoided, often leads to the conclusion that much published research is trivial and slipshod. This is, in fact, stated by some of its proponents, for example, Caplow and McGee: 'The multiplication of specious or trivial research has some tendency to contaminate the academic atmosphere and to bring knowledge itself into disrepute. The empty rituals of research come to be practised with particular zeal in unsuitable fields. . . .'[1]

In England, the desire to expand the universities, and the consequent need to train less well prepared students with less well prepared teachers, has resulted in criticism of the system of the control of university scientists by the scientific community. Sir Eric Ashby has expressed some typical sentiments: '[The] young man who is inspired to devote his career to the real purpose of a university, which is teaching at the frontiers of knowledge, finds himself obliged to enter a different career: the rat-race to publish. And to publish what? It must be "original": miniscule analyses of kitchen accounts in a medieval convent; the structure of beetles' wings – some beetle whose wings have not been studied before; the domestic life of an obscure Victorian poet; the respiration cycle of duckweeds. All, no doubt, interesting; all, in a way, at the frontiers of knowledge, even though it is crawling along the frontier with a hand-lens; all original, in the sense that no one has done them before; but all (with some few exceptions) so secondary to the prime purpose of a university'.[2] Such sentiments may be true. The problems of the quality of teaching will not be discussed here, but another implication of the two preceding quotations should be emphasised, namely, that the contractual theory of the organisation of science accounts not so much for its *organisation* as for its *disorganisation*. . . .

The Organised Disorganisation of Science
This chapter has been concerned with the problem of the co-ordination of scientific efforts when such co-ordination requires continuing close work involving two or more individuals. First described were the traditional solutions to the problem, whereby scientists collaborate freely. They work together to solve a scientific problem that interests both of them, together prepare publications reporting the solution, and share equally in the recognition given to the work. This type of joint effort is altogether compatible with the theory of control described in earlier chapters – an exchange of information for recognition.

Traditionally, scientists also obtain assistance from their students. When this occurs, the university scientist is primarily responsible for the information contributed to the scientific community and receives most of the recognition given the work, although he may share it with his students, especially if he competes with his colleagues for students. The dependence of university scientists on graduate students may lead them to avoid risky and long-term projects, and, when teamwork with students occurs, free collaboration with other professionals may become more difficult. These are slight disadvantages to teamwork in academic settings, however. (In industrial and governmental research, risk-taking is also inhibited because it jeopardises the status of the leaders of more or less permanent research groups.)

Other factors are more important in the tendency of modern science to supplement or supplant traditional forms of co-ordination with more complex forms. Much modern research is characterised by scarce and expensive facilities that must be shared by many scientists. This separates the scientist from the means of production: others determine the uses he will make of research tools. The scientist must be oriented to those in charge of research facilities and must persuade them to give him access to the facilities. His readiness to respond to such officials may reduce his readiness to respond to the community of scientists.

In addition, modern research often involves more specialised skills than any single individual can command. Research is often made more effective if some highly trained individuals become technicians, placing their skills at the disposal of others.

Like other professions, then, modern science is characterised by the splitting of the professional role into the roles of the administrator and the technician. Leaders necessarily become politicised, oriented to obtaining funds and access to facilities and co-ordinating the efforts of others. Technicians become means-oriented, interested in performing their specialised skills for extrinsic rewards and uninterested in the recognition given by the scientific community for the attainment of scientific goals.

If this occurs, the information-recognition exchange theory of scientific organisation is no longer applicable. Control is exercised by hierarchical authority within research groups and by political powers outside of them. Scientists become more interested in their particular organisations and in the reactions of politically powerful leaders than in the opinions of the wider scientific community. Consequently, the complex organisation of science leads to disorganisation – disorganisation in terms of the information-recognition exchange theory of organisation. The organisation that may be supplanting the traditional information-recognition exchange will probably be ineffective for achieving scientific goals. Basic research in complex organisation tends to become ritualistic, reflects low commitment to scientific values, and is displaced easily by practical goals. Attempts are made to accommodate the non-scientific donors of research funds, and the technician mentality sees little difference between pure and applied research in any case. When recognition from the scientific community loses its value, recognition will be sought from other users of research. As the norms of independence for professional scientists become abridged, scientists will come to feel less responsible for achieving scientific goals.

Notes
1. T. Caplow and R. McGee, *The Academic Marketplace* (Basic Books, New York 1958), p. 221.
2. E. Ashby, *The Listener* (1 June 1961).

APPENDIX 6

The Occupation of Science

N. D. ELLIS

From *Technology and Society* vol. 5 (1969), p. 33

The large-scale employment of scientists and technologists in government and in industry is a recent phenomenon. Most of the laboratories which are now an accepted feature of modern government and industry, have been established only since the beginning of this century. As Krohn[1] remarks, the centre of gravity of science has shifted radically. The support of research by government and industry has grown to enormous proportions, and correspondingly, the traditional bases of support – university funds and private philanthropy – have declined in significance. Several writers have argued that this change of location has been accompanied by certain attitude changes. Increasingly, the large-scale formal organisation has come to be accepted as the appropriate environment for research, and the lone independent investigator has lost his previous importance. Emphasis has shifted from the independent intellectual, pursuing knowledge in a context of maximum freedom, to the team directed along profitable lines. Also, research is increasingly justified (even by the universities) by its potential value in the achievement of human welfare, rather than by the value of knowledge for its own sake.

Some statistics from a recent government survey illustrate the extent to which science and technology are carried on in the industrial context (including both the nationalised and the private sectors). Data from this show that during the period 1964–5, 63·9 per cent of the total research and development effort of this country (measured in terms of cost) was carried out in the industrial sector and accordingly, approximately 60 per cent of the total man-power was deployed in this sector. During this period, the overall contribution of the universities to the total effort was only 6·5 per cent. Current expenditure by type of research activity was distributed thus:

12·6 per cent of the total expenditure was on basic research.
26·0 per cent of the total expenditure was on applied research.
61·4 per cent of the total expenditure was on development work.

Within the category 'basic research', 44 per cent of the effort was concentrated in the universities.

On the basis of this evidence, it is reasonable to suppose that the 'typical' work situation of the contemporary scientist (or technologist) is the industrial laboratory, and that his work is probably applied research and/or development work.

The knowledge – provided by historians and sociologists – we already have about the nature of the 'occupation of science' seldom reflects this contemporary reality. Historians have tended to concentrate their efforts upon problems and events occurring in past centuries. We know a considerable amount about the early proceedings of the Royal Society and other esteemed scientific institutions of the past; yet little is known about the origins and development of the industrial laboratory – the most typically modern institution of science. Likewise, sociologists have debated at length about the nature of the social organisation of 'pure' science, devoting much of their effort to defining its distinctive ethos. It is therefore a relatively simple matter to read about the activities and thoughts of outstanding scientists of the past and the sentiments of pure scientists today. However, if we wish to consider the large majority of scientists and the largest sector of modern science – industrial scientists and technologists and their work – the literature is, as yet, limited. Very little is known about the present situation of the industrial scientist in Britain, and about how this has changed during this century.

Historians and sociologists have tended to avoid problems of this kind by excluding applied science, and those engaged upon it, from their discussions. Their definitions of 'science' and the 'scientist' are often based upon an assumption that 'pure' research constitutes 'real' science, and that the academic environment is the most natural one for the scientists. Other kinds of knowledge-seeking activities based upon the established scientific and technological disciplines are seen as being, at their best, poor substitutes for the real thing. These writers – themselves academic researchers – seem to have adopted a *Weltanschauung* frequently associated with the 'world of learning' in their own studies of science. The belief that the most worthy vocation is the pursuit of knowledge removed from any economic and social context is still widely held by academic scientists; my own survey provided evidence of this. Associated with this belief is the view that the use of the knowledge and skills of a scientific discipline for reasons other than the furtherance of that discipline, is 'corrupting' and often involves second-rate work. The preoccupations of many sociologists in this field appear to reflect the self-consciousness of the academic rather than an inclination to examine situations, and individuals' perceptions of these, which are radically different from his own.

Broadly speaking, previous studies in this field have often approached the problem from the wrong end. Many writers seem to have started out by thoroughly examining their own academic environment, and have then inquired how scientists can survive outside it. It has been widely assumed that many industrial scientists are 'itching' for an opportunity to do 'real' science – pure research – and that creativity is seriously restricted in an environment which does not provide opportunities for work of this kind. Thus the industrial scientist has often been portrayed as the 'aspiring academic', who has through some misfortune, wandered into the wrong camp. Evidence from recent studies and my own survey suggests that this view is very far from the real situation. The large majority of scientists whom I interviewed expressed a definite preference for applied research and did not regret being unable to pursue fundamental knowledge in their fields. . . .

My own discussion involves rather different definitions of 'science' and the 'scientist' from those mentioned above. Applied research and development work are included under the rubric 'science', as is 'pure' research. Accordingly, those engaged upon applied science have as equal a right to the title 'scientist' as has the academic. If these other kinds of research are included, then the traditional distinction between scientist and technologist loses much of its supposed significance. In many of the laboratories I visited a distinction between the scientist and the technologist, according to the subject of the degree, had very little meaning; graduates of the 'pure' sciences and of technologies were often deployed on the same problems in multi-disciplinary research teams. Thus the term 'industrial science' may be taken as referring to the whole spectrum of research and variety of disciplines found in this sector.

The Orthodox Model of the Industrial Scientist

Many previous studies of the industrial scientist have taken as their central theme the conflict between two cultures – that of 'management' and that of 'science'. These writers have assumed that this conflict characterises many of the problems which occur in the administration of the industrial research laboratory. These two cultures – or 'ideologies' as they are sometimes termed – imply opposing sentiments about what is valuable and worth while. . . .

Kornhauser attempts a sociological explanation of this phenomenon. His model is based on the premise that the industrial scientist identifies with a 'professional' colleague community external to the organisation in which he works; this involves an 'ideological' commitment on the part of the scientist which conflicts with the 'bureaucratic ideology' of the management. . . .

Kornhauser develops his theme by examining the mechanics of this conflict; he describes four spheres in which it can occur.

Goals

'Strains between science and organisations emerge first of all in the formulation of research goals.'[2]

The 'professional community' and management have very different views about what is important and worthwhile scientific research. The scientist – being a member of this community – favours basic research goals, free from the constraints imposed by commercial objectives. 'Scientific institutions seek to instil in their members a commitment to the growth of science as a discipline of the human mind, and an obligation to sustain the integrity of science itself. Individual scientists are not equally socialised to the values and norms of science any more than they are equally well trained in the technical skills of science. Nevertheless, scientists generally bring to their work an internalised set of standards for scientific activity. By virtue of their training they expect to contribute to science in their sphere of special competence.'[3] This statement contains the supposition that industrial scientists have generally internalised the values of 'pure' science during their undergraduate training. In addition, it is further assumed that these values and beliefs are sustained in the industrial context by the 'professional community'.

Scientists and managers come to have very different interests in research; management's goals for the laboratory differ from those its participants bring to it. A quotation from Kornhauser's text succinctly describes this dilemma: '. . . the issue of basic versus applied research expresses the underlying tension between professional science and industrial organisation. Professional science favours contributions to knowledge rather than to profits; high-quality research rather than low-cost research; long-range programmes rather than short-term results; and so on. Industrial organisation favours research services to operations and commercial development of research. These differences breed conflict of values and goals; they also engender conflicting responsibilities and struggles for power.'[4]

Controls

Conflict in this sphere stems from the interaction – in the context of the industrial laboratory – of two conflicting principles of control, one based on the principle of hierarchy and the other upon that of colleagueship. In the bureaucratic organisation – i.e., the research laboratory – authority is vested in a series of offices that are hierarchically ordered. In contrast with this, 'professional' control is exercised by a company of equals. Thus, whereas scientific institutions are based on various kinds of colleague authority, government and industry rely primarily on administrative responsibility.

Incentives

Since the 'worlds' of learning and commerce hold very different views about what is valuable and important, each will offer its own kind of incentives for the industrial scientist. 'The scientific profession seeks contributions to knowledge by soliciting research papers for professional meetings and journals, and by rewarding intellectual excellence with honours and esteem. The industrial firm seeks contributions to production and sales by soliciting new or improved devices, and by rewarding commercial success with promotions in a hierarchy of status, income and authority.'[5]

Influence

Kornhauser examines several sources of resistance to the utilisation of research findings and the influence of research personnel. The difference between the *Weltanschauung* of the scientist and that of management and other groups in the parent organisation may contribute to the isolation of the laboratory. Barriers to communication between research staff and other groups in the organisation may result from a 'vicious circle' in which all parties adopt the view that a minimal level of contact is the happiest solution for all concerned. Consequently the *Weltanschauung* of each party will become firmly established and both sides come to define each other as 'hostile' and 'alien'.

Some Limitations of this Model

This model certainly expresses the contrasts between those sentiments which characterise 'pure' or 'academic' science and those of industrial management. The objectives of industrial science are ultimately 'commercial'; the support given to research is motivated by a belief that it can contribute towards the achievement of certain economic goals. Thus, any scientific or technological advance made by the research staff is finally subjected by management to an evaluation according to economic criteria. There is an enormous difference between this 'pragmatic' problem-solving activity and that usually associated with academic research. Both the initial conceptualisation of the problem, and the criteria of adequacy applied in the evaluation of a solution are very different.

This model does, however, go much further than this. It does not simply assert the truism that there is a vast 'ideological' or 'cultural' gap between the 'worlds' of 'pure science' and 'commerce', but also that the majority of industrial scientists are integrated into the former, and are alienated from the latter. Several suppositions are used to justify this maxim; these have been investigated in some recent studies and my own survey.

The assumption that a significant proportion of science students come to internalise the 'ethos of pure science' during their undergraduate training has already been radically modified by findings from a study by Box and Cotgrove.[6] Their evidence shows that a significant proportion of science students have an instrumental view of science; they acquire the knowledge and skills of science and do not seek public recognition as scientists, being prepared to use their expertise for occupational advancement and to abandon it when expedient. Evidence from another study, that of Avery,[7] shows that even if the young graduate carries the 'pure science ethos' with him to the industrial laboratory, this is soon discarded. A process of enculturation occurs during which the scientist learns a new culture and comes to adopt it as a necessary ingredient of his working life.

The Table[8] provides comparable indices of 'importance to overall work satisfaction' for a series of items. High scores on items C, E, H, I, and L are associated with a 'pure science ethos'. The high scores for those items obtained from the sample of academic scientists certainly corroborates this. Academics, judging from both their scores on this battery of items and their answers to questions in the interview, undoubtedly attach considerable importance to being free to choose their own research problems and thus do fundamental research. In contrast to this, industrial scientists seldom emphasised the importance of his kind of freedom. Their scores on these items (C, E, H, I, L), together with their answers to interview questions, clearly showed that very few industrial scientists embraced the 'pure science ethos', irrespective of whether they may have held these sentiments when they left the university. Most of those interviewed expressed a definite preference for applied research; the reason most frequently given for this was that applied science provided more intrinsically satisfying results – a tangible end product – than those usually associated with basic or fundamental research.

Notes
1. R. G. Krohn, *Institute of Radio Engineers Transactions on Engineering Management* vol. 8 (1961), p. 133
2. W. Kornhauser, *Scientists in Industry* (University of California Press, Berkeley, California 1962), p. 16.
3. ibid., p. 18.
4. ibid., p. 25.
5. ibid., p. 156.
6. S. Box and S. Cotgrove, *Brit. J. Sociol.* vol. 17 (1966), p. 20.
7. R. W. Avery, *Institute of Radio Engineers Transactions on Engineering Management* vol. 7 (1960), p. 20.
8. See Table 2, p. 89.

The Flow into Employment of Scientists, Engineers and Technologists

Committee on Manpower Resources for Science and Technology

From the Swann Report, *The Flow into Employment of Scientists, Engineers and Technologists* (Cmnd 3760, HMSO 1968), pp. 2–3 and 91–3

Outline of the Argument

As we shall show, there are a number of serious imbalances in the present patterns of flow of qualified manpower compared with the needs of employment. Indeed in that there is a concentration of scientific talent in the fundamental research sector (particularly in universities) and a very significant movement abroad, with a consequent starving of industry and schools, the figures reveal a positively dangerous situation.

We do not think that the falling proportion of pupils following science in schools is solely due to the failure of sufficient good graduates to enter teaching; or that the difficulties facing British industry in competing with the United States, Japan, or various European countries result only from the limited extent to which it has been able to recruit and use effectively the most able graduates. Dr Dainton's report[1] pinpoints other important factors in the swing away from science: and we have often heard it said that the failure of the best science graduates (and, to some extent, of the best technology and engineering graduates) to go into British industry in sufficient numbers is not so much the *cause* as the *result* of the present state of much of that industry. Nevertheless, it is clear to us that the situation in the schools and in industry cannot be satisfactory unless more of the best graduates enter these sectors. Altering the patterns of the flow of qualified manpower will not put everything to rights, but equally, things will not be right until these patterns are altered.

The diagnosis, let alone the cure, of the malaise that affects the British economy is clearly far beyond our terms of reference. The Report of the Working Group on Migration[2] has recently displayed some symptoms of this malaise, and other Groups are examining particular aspects of it. We have therefore confined our attention largely, though not entirely, to the task of bringing about improvements through the education system and particularly through higher education. Our recommendations, summarised in Chapter XII, attack only one corner of a vastly complicated situation. The measures we propose for education, or indeed any others, will not in themselves suffice; they depend for their success on complementary developments in employment. Only through joint action by education and employers, especially in industry, can progress be made in solving present manpower problems.

Nevertheless we propose that education take the initiative in dealing with these problems. It should be able to respond more readily (despite the time taken for students to qualify and enter employment) than the multiplicity of small firms that characterise the structure of much of our industry; and at the postgraduate level the prospect of rapid change in the pattern of new supply. Progress with the problems of deployment and utilisation of qualified manpower already in employment (which are being studied by other Groups) is equally important but may well take longer to achieve.

The short-term measures we propose, to meet the immediate needs of industry and the schools, relate to postgraduate training and awards policy, and to means of alleviating the shortage of science and mathematics teachers. They are all matters on which the universities (where postgraduate work is most strongly developed) or Government can give a rapid lead.

We also propose some longer-term reforms. The entire process of education and training for a career takes a very long time. It is necessary therefore to consider, not the pattern of demand today, but the likely pattern of demand ten or fifteen years ahead, when students now in the upper reaches of school, or entering higher education, will hold responsible positions in employment. Such forward looks are inevitably speculative, but it is useless to delay decision in the hope that things will become clearer, for by then it will be too late. We have therefore attempted to examine the experience and practice of the most highly developed sectors of industrial employment, both in this country, and more especially in the United States. We have tried to see how such advanced industrial sectors deal with the problems of rapid technological change, and we believe that we can, as a result, identify the types of graduate that advanced industries will want a decade or more hence. From this we attempt to assess the implications for educational and manpower policy today.

The day is past when universities were for the elite only. With a rising proportion of the population entering higher education, employers will increasingly have to seek able new recruits from this source, to fill an ever-widening range of jobs. Moreover we see the study of science and technology as a desirable preparation for an increasingly wide spectrum of occupations extending well beyond

160 / SCIENCE OBSERVED

the traditionally vocational employment in these subjects and into fields outside science and technology as such. This is, essentially, the theme of the measures we propose for reform in the longer term. . . .

SUMMARY OF RECOMMENDATIONS

Short-term Measures

Postgraduate Studies and Industry

1. We agree with the guidelines of the University Grants Committee for growth in the number of postgraduate places in science, engineering and technology during the present quinquennium; the Committee's provision related to a total of nearly 18,000 places in 1971–2. We hope that, within these guidelines, the numbers embarking on postgraduate work, whether immediately after graduation or following a period in employment, will keep in step with the numbers graduating – though this may mean considerable stringency on the numbers proceeding to research training. Policy for postgraduate studies and awards should be guided by the need to match these more closely to the requirements of employment, especially in industry.

2. There should therefore be a change of emphasis towards shorter periods of postgraduate study and the trend from research to advanced course work should be accelerated. The proportion of students following postgraduate studies after a period in employment (that is, post-experience students) should increase rapidly; and flexible course arrangements should be developed to make it easier for persons in employment to attend. The content of postgraduate work should be reviewed; new approaches to postgraduate training should be encouraged.

3. The universities should examine the nature and purpose of the Ph.D. from first principles and consider drastic action to bring within its scope other forms of postgraduate training more closely orientated to the requirements of industry.

4. Industry should vigorously recruit persons qualified in science, engineering and technology, especially the ablest graduates in these fields; should see that they are fully and effectively employed, and that attractive and challenging careers are open to them. Industry must give clear guidance to educationalists on its requirements and, using to the full the provisions of the Industrial Training Act, become intimately involved in the planning, conduct and support of postgraduate and post-experience education and training.

5. Employers in industry should seek to develop steady policies for research and development and for the recruitment and employ-

ment of scientists, engineers and technologists, taking full advantage of relevant Government measures such as tax allowances and investment grants.

6. The manpower needs of industry should be taken into account when the pay of scientists, engineers and technologists is considered; in applying the principle of 'fair comparisons' in reviews of the pay of scientists, engineers and technologists, regard should be had to the inherently greater freedom and security in university and public service employment.

7. Scientists, engineers and technologists should be able to move readily between the main sectors of employment – universities, schools, industry, Government. Every attempt should be made to remove artificial administrative barriers to mobility; in particular the problem of transferability of pension arrangements should be further examined.

Recruitment of School Teachers

8. So long as the present shortage of qualified manpower to teach science and mathematics continues, there should be established among teachers 'priority categories' in these subjects to whom preferential salaries would be paid; there should be 'merit additions' to salary for approved higher degrees in science and technology, and financial recognition of additional responsibilities in science teaching, such as laboratory organisation.

9. The possibilities of recruiting to teaching mature and able scientists, engineers and technologists from industry and Government should be carefully examined. There should be greater recognition, by way of salary increments, for appropriate industrial experience. Local education authorities should consider re-employing specialist teachers in science and mathematics who have recently retired.

10. More graduates should gain some first-hand acquaintance with teaching during their university career. Attempts to introduce into the first degree course in science, engineering and technology some knowledge and practical experience of school teaching should be encouraged. There should be further experiment in combining study in a major field of science, engineering or technology with preparation for teaching, without lengthening the first degree course.

11. Ways should be explored of bringing to bear the resources of universities on the earlier stages of science teaching in schools, for example, by providing 'locum tenens' services for teachers wanting to attend courses, or through extension lectures.

Measures for the Longer Term

12. To maintain the growth in the numbers of scientists, engineers

and technologists who qualify each year, universities should make every effort to accommodate students to study for first degrees in these subjects.

13. To meet current and future needs of employment, and to give students of science, engineering and technology some understanding of the society in which they will work, universities should consider making the first degree course in science, engineering and technology broad in character, through multi-disciplinary approaches to these subjects and by introducing relevant study in other fields such as economics, sociology, law, etc.

14. To prepare for and assist in this change there should be a detailed study of current curricula in science, engineering and technology in university education and of the balance between specialised and more general studies, in relation to career needs. We suggest that the investigation should be initiated by the Committee of Vice-Chancellors and Principals in conjunction with the University Grants Committee.

15. Institutions of higher education have a particular contribution to make to continuing education and training throughout the individual's career. In due course universities should come to regard as part of their normal provision post-experience courses for mature scientists, engineers and technologists. Employers should be closely involved in the mounting of post-experience courses so that successful completion is linked with career prospects.

16. The flow of scientists and technologists into employment is likely to present continuing problems which must receive further study. Progress on the measures taken to deal with these problems should be regularly reviewed; to assist this, statistical studies should be refined and extended. The factors shaping students' aspirations to postgraduate work, and their career decisions, should be studied. The interaction of sources of support for postgraduate work should be examined. Techniques for forecasting longer-term trends in the needs for scientists and technologists should be further developed.

Notes
1. Report of the Enquiry into the Flow of Candidates in Science and Technology into Higher Education (Cmnd 3541, HMSO, London 1968).
2. *The Brain Drain*: Report of the Working Group on Migration (Cmnd 3417, HMSO, London 1968).

APPENDIX 8

An Attempt to Quantify the Economic Benefits of Scientific Research

I. C. R. BYATT AND A. V. COHEN

From *Science Policy Studies No. 4: An Attempt to Quantify the Economic Benefits of Scientific Research* (HMSO 1969), paras. 2, 14, 23–6

The benefits of scientific research (including applied science) may be summarised as follows:

Manpower Benefits
(a) The output from the educational system of graduates familiar with recent scientific advances and taught by those pursuing research, including the benefits of there being diffused throughout productive society a significant body of persons familiar with recent scientific advances. Those graduates remaining within the educational system, whether in universities, colleges or schools, play an essential part in this process.

(b) As an addition to (a), the output of persons with higher degrees and diplomas from the educational system and their consequent employment.

Research Benefits
(c) The cash benefits to industry or others from exploiting 'mission-oriented' or 'applied' research, that is, research in fields whose application is evident, for example, work in agriculture and medicine. An incidental additional benefit is technological spin-off, as it is normally understood.

(d) Other material benefits from research such as described in (c) · for example, in medicine the working time gained through lower absenteeism and morbidity, and savings on medical expenses.

(e) Humanitarian and social benefits associated with (c) – e.g. in medical research, greater longevity.

(f) Applications in industry, often at a much later date, of basic ideas discovered during 'curiosity-oriented' pure research. These ideas tend to arise unpredictably and, when applied, to give rise to big industries, or to complete reorientation of existing industries, perhaps several decades later, though it is often maintained that this period is shortening.

(g) The absorption into the infrastructure of science, and the subsequent industrial application, of a whole host of apparently minor discoveries, or of general principles of no apparent immediate relevance made during the course of 'curiosity-oriented' research.

(h) Benefits associated with (f) arising from 'curiosity-oriented' work conducted in the United Kingdom simultaneously with, and similar to, that giving rise to major scientific discoveries abroad, since this enables United Kingdom research workers quickly to appreciate what has been discovered. A key scientific discovery recently made might be related to earlier discoveries and ultimate commercial application accelerated, perhaps through the mechanisms suggested in (i).

(i) Further benefits associated with (f) since those who have at least some recent research experience and are familiar with the latest

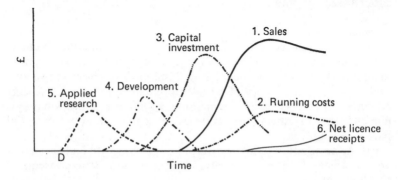

scientific developments are particularly well equipped to be employed to look for the possible commercial developments of scientific discoveries, whether made at home or abroad. This benefit is distinct from that measured by the output of trained graduates, (a) and (b), since these people will need to keep in contact with active British basic researchers, not only to maintain their scientific vitality, but also in order to become aware of the latest scientific discoveries wherever made.

(j) The cultural values of the activity. . . .

The economic value of the research derives from the total economic surplus – i.e. the total economic benefit derived from the final product or process, less the costs of applied research, development, investment and manufacturing. This surplus is a wider concept than the profit of an individual firm. It includes the profits of firms, and indeed the money, costs and receipts of manufacturing firms may, in a number of cases, be good measures of the economic benefit to society as a whole. (The figure shows a typical pattern of cash flow for a firm.) In most cases, however, economic benefits will accrue to others. Final consumers will gain if an innovation reduces the cost of the goods and services they buy. The profits of other firms may rise or fall. For example, the introduction of nuclear energy affected the manufacturers of conventional generating plant, the Central Electricity Generating Board, the Area Boards, the National Coal Board, the oil companies, and both industrial and domestic buyers of electricity. Thus it would not be sufficient to confine attention to the accounts of the Atomic Energy Authority and the nuclear consortia. The effect of an innovation on the whole economic system may be considerable; for example, the costs of moving redundant coal miners to areas of greater employment opportunities should, on this basis, be included in the costs of nuclear power. Many, but not all, of these costs and benefits may show up in the accounts of firms either directly or indirectly affected by an innovation. Some, like the economic costs associated with atmospheric pollution, or the saving of leisure time, may not be shown in any set of existing cash transactions. When examining the benefits for a particular country it will probably be found that the relative magnitude of benefits and costs will differ considerably between different countries, because of differences between national goals, which may be either political or economic. Different economic goals, for example employment or balance of payments considerations, may be dealt with by using 'shadow prices'. If the innovation produces a major disturbance to the economic system, significant problems of analysis arise. In principle, it is necessary to trace out a whole series of effects, some of which may be non-marginal. In doing this, one should take into account all relevant systems effects, including what are variously known as externalities, spill-overs and neighbourhood effects. . . .

Much scientific work is linked not with only one product, industry or process, but with several. Similarly new products, etc. often draw on a number of scientific discoveries. Discoveries go into a pool of ideas which are subsequently drawn on in a whole range of ways. Some discoveries spark off other discoveries in quite unrelated ways. Hence much research cannot be allocated to particular new products, etc. nor can new products be allocated to specific pieces of research. This makes analysis difficult, although not impossible. It is still possible to attempt to analyse the effects of hypothetical marginal changes in research expenditure throughout this system, even though the several parts of the system cannot be disentangled.

A possible way to analyse the returns to research, whether to the world as a whole or the United Kingdom and thus to arrive at an idea of what level of expenditure would be economically justified, is to postulate small adjustments in this complex system and to try and trace out their consequences. The effects of adjusting research expenditure in any given time period may then be analysed by postulating small adjustments in this variable and tracing out the effect on the economic value of the research.

Hypothetical adjustments in research expenditure would have slowed down (or accelerated) the exploitation of new products and new processes, either directly or by slowing down or accelerating the making of other scientific discoveries which have this effect. They will not usually have prevented innovation indefinitely, as, if a research project is abandoned, someone else, either later in time or in another laboratory, which may be in another country, will probably make the discovery.

The consequential slowing down of the exploitation of new products and processes would usually result in an economic loss and an acceleration of the exploitation will usually result in economic gain. The extent of the *change* in the net economic benefit to society, when appropriately discounted, can be compared with that change in expenditure on research which results in this change. Only if the increase in net economic benefit from all the industries associated with the scientific discovery, when discounted, is equal to or exceeds the increased expenditure on research, will the extra expenditure have been worth while.

APPENDIX 9

The Value of Curiosity-Oriented Research

M. GIBBONS, J. R. GREER, F. R. JEVONS, J. LANGRISH
AND D. S. WATKINS

From *Nature* vol. 225 (1970), p. 1005

A way to approach the problem of quantifying the economic benefits from curiosity-oriented research has recently been suggested by Byatt and Cohen.[1] It consists essentially of identifying key discoveries which have had profitable applications and then estimating the economic effects of notional marginal delays in the timing of these discoveries; that is, of attempting to assess how much less wealth, suitably discounted to a common year, would have arisen if, because of a smaller scale of effort in particular areas of research, certain discoveries had been made later than they actually were.

We report in this article some studies which formed part of a project to assess the feasibility of this approach by trying to apply it to some recent examples of technological innovation. The examples were chosen after due consultation with the Working Group on the Economic Benefits of Scientific Research which has been established by the Council for Scientific Policy. The principal areas selected were the Chorleywood bread process, the float glass process and cryogenics. Some other areas of innovation were also considered more briefly.

The term 'curiosity-oriented' used by Byatt and Cohen is clearer than 'pure' or 'basic', indicating research justified by curiosity about topics for which no application is apparent. According to Byatt and Cohen (benefit (f) of their article, paragraph 2), ideas discovered during curiosity-oriented research, which 'tend to arise unpredictably', are sometimes applied later in industry. This is contrasted with 'mission-oriented' research, defined by Byatt and Cohen (benefit (c)) as 'research in fields whose application is evident'.

Thus work cannot properly be described as curiosity-oriented if it can be shown that some application was envisaged; though naturally this does not imply that scientists necessarily lose interest just because there is a practical goal in view. The term 'basic' is commonly used in a wider sense to include fundamental investigations of topics with a quite obvious application.

These distinctions are not just semantic subtleties. They are relevant to the possible policy problem of the extent to which basic research should be funded regardless of whether social or economic benefits seem relatively likely. The alternative is to try to identify topics which have obvious social and economic value and selectively support work in them, even though this work may be quite basic.

Chorleywood Bread Process
The principal feature of this process[2] is the replacement of a lengthy period of dough 'development' involving fermentation by a combination of a controlled amount of intense mechanical work and the use of certain chemicals known as 'improvers'.

The effect of mechanical work was described by mission-oriented researchers[3] in the United States in 1926, thirty-five years before the Chorleywood process was announced by the British Baking Industries Research Association (BBIRA). No commercial application resulted until J. C. Baker in the United States was able to combine the effects of mechanical work with the use of improvers. Baker carried out careful systematic studies of a mission-oriented nature and no large input of curiosity-oriented research was involved.[4]

Baker's 'Do-Maker' process was not suitable for the needs of the British bakers or housewives, but the Chorleywood process is now used by more than two-thirds of the British baking industry. The Chorleywood process uses ascorbic acid (vitamin C) as a fast acting improver in addition to conventional improvers which were discovered in the 1920s by accident and by empirical research. The commercial availability of ascorbic acid depends on research in organic chemistry carried out in the 1930s but, had ascorbic acid not been available, other fast acting improvers could have been used.[5]

The technical knowledge forming the background to the research which produced the Chorleywood process was available in the 1930s, but could not have been used in Britain before the 1950s when chemical additives in flour and bread were allowed after studies by BBIRA.

While the allocation of greater resources to BBIRA might conceivably have brought about the development of the Chorleywood process slightly earlier, there is nothing to suggest that greater curiosity-oriented research effort would have had any accelerating effect.

Float Glass
The float glass[6] process seems to be a typical 'technological discovery' in that it cannot be described in terms of the application of a specific discovery made by academic or other curiosity-oriented scientists. The concept of using molten tin for the manufacture of flat glass was patented in America in 1902,[7] Pilkington had to spend seven years

APPENDIX 9 / 169

and £4 million on development work, however, before announcing
the new process in 1959.

After the process had been developed empirically to the stage of
producing satisfactory glass, scientists became involved to increase
the understanding of the process and to solve technological problems;
for example, the cause of a surface bloom was traced to a layer of tin
dissolved in the glass. Such work made use of certain fundamental
concepts and analytical techniques which may have derived from
curiosity-oriented research; but it does not seem possible to apply
the concept of marginal delay to these connections between science
and technology, which occurred after the initial development of the
process.

Cryogenics

We chose cryogenics to complement bread and glass manufacture,
because it seemed *a priori* more likely to show dependence on
curiosity-oriented research. We concentrated on the cryogenic gas-
separation industry because it is too early to assess the economic
impact of work at temperatures below about 20 K.

The history of low temperature research yielded one case that
seemed at first to lend itself to Byatt–Cohen analysis. The Joule–
Thomson effect,[8] discovered in 1852, was in 1895 made the basis of a
gas liquefaction process by Linde[9] in Germany and independently by
Hampson in England.[10-12] It is at least plausible to argue, however,
that a delay in the discovery might have accelerated rather than
delayed its application. The Joule–Thomson effect was assimilated
into science as a small-scale effect and it required considerable
insight on the part of Hampson and Linde to see it, as it were, 'out of
context' as part of a powerful industrial process. Had it been discovered after the failure of the Pictet[13] and Cailletet[14] approaches to
liquefy oxygen on any but the very small scale, its potentialities as
the basis of an industrial process might have been realised somewhat
earlier.

Much of the explanation for the timing of the development of the
industrial process seems to lie in market rather than technical
factors: the realisation of a large potential market for oxygen, to
meet which only a relatively costly chemical process was available
(ref. 15 and personal communications from personnel of the British
Oxygen Co., formerly Brins Oxygen Co.).

Other Innovations

Of several other major innovations considered more briefly in the
course of this project, it seems that only nuclear power and silicones
can legitimately be considered to be based on specific discoveries of

curiosity-oriented research. In those two cases, the important influence of wartime development work makes it difficult to apply the concept of notional marginal delay; the timing of the innovations may have been determined more by the Second World War than by the timing of the curiosity-oriented discoveries. The massive effort put into the development of an atomic bomb is well known. Silicones could be said to depend on specific curiosity-oriented discoveries made by Kipping in 1904 and 1908. Silicones, however, were first produced commercially in 1943 by Dow–Corning to satisfy military needs and post-war production was based not on one of Kipping's reactions but on a synthesis developed by Rochow working for General Electric.[16]

In several other cases, however, curiosity-oriented research may have been involved in the sense that knowledge and techniques were passed on through the educational process and made available for the solution of industrial problems via the supply of scientifically trained manpower. An example of such a problem is given in connection with float glass.

In a further search for innovations which could be directly-linked to curiosity-oriented research, issues of the *New Scientist* for the period January to June 1968 were surveyed, concentrating particularly on the Science in Industry feature. Of a sample of seventy-four British innovations thus obtained, fifteen (20 per cent) were considered possibly susceptible to Byatt–Cohen analysis, but in fourteen of these the chief reason for categorising them thus was lack of precise knowledge.

Byatt–Cohen Hard to Find
From the studies we have summarised, we conclude that Byatt–Cohen type innovations are quite difficult to find, let alone investigate. Interactions between science and technology are usually too complex for the method of notional marginal delay. Only rarely is it possible to pinpoint specific curiosity-oriented discoveries from which wealth-producing applications are derived. Even when this can be done, other factors affecting the timing of innovations, such as market factors or wartime pressures, obscure the effects of possible delays in discoveries.

It should be emphasised that we do not conclude that curiosity-oriented research is useless in economic terms. On the contrary, in view of the importance of the problem, we feel that further work should be directed to exploring various other avenues through which curiosity-oriented research may lead to economic benefits.

Notes
1. I. C. R. Byatt and A. V. Cohen, *An Attempt to Quantify the Economic*

Benefits of Scientific Research, Science Policy Studies No. 4 (HMSO, London 1969).
2. G. A. H. Elton, *Proc. Roy. Austral. Chem. Inst.* vol. 32 (1965), p. 25.
3. C. O. Swanson and E. B. Working, *Cereal Chem.* vol. 3 (1926), p. 65.
4. J. C. Baker, *Bakers Weekly* vol. 161, no. 11 (1954), p. 60.
5. H. Jorgensen, *US Patent* 2 (1939), pp. 149, 682.
6. L. A. B. Pilkington, *The Glass Industry* vol. 44, no. 2 (1963), p. 80.
7. W. E. Heal, *US Patent* 710 (1902), p. 357.
8. J. P. Joule and W. Thomson, *Phil. Mag.* series 4, vol. 4 (1852), p. 481.
9. R. Linde, *German Patent* 88 (1895), p. 824.
10. W. Hampson, *J. Soc. Chem. Ind.* (1898), p. 411.
11. W. Hampson, *Industrial Gases (Quarterly)* (Sept. 1921), and others.
12. W. Hampson, *UK Patent* 10 (1895), p. 165.
13. R. Pictet, *CR Acad. Sci.* vol. 85, pp. 1214, 1220.
14. L. Cailletet, *Comp. Rend.* vol. 85 (1877), p. 815.
15. Monopolies and Restrictive Practices Commission, *Report on Supply of Certain Industrial and Medical Gases* (HMSO, London 1965).
16. E. G. Rochow, *Chemistry of the Silicones*, 2nd edn (Wiley, New York 1950).

APPENDIX 10

Conditions for Successful Innovation

Central Advisory Council for Science and Technology

From *Technological Innovation in Britain* (HMSO 1968), pp. 3–5

Any firm, or indeed any country, engaged in world trade in advanced industrial products, must repeatedly modernise its manufacturing processes and introduce new or up-dated products if it is not to lose markets and go out of business because of competition from advances elsewhere. Hence the constant need for market awareness and for technological innovation. The latter, however, is an extremely risky activity. The history of technology is studded with examples of commercial failures; of inventions not taken up; of projects cancelled through daunting escalation of costs; of innovations pre-empted by faster rivals; and of new products spurned by an unprepared and unreceptive market.

Of the various factors which make for success in technological innovation, five stand out particularly. They are:

(a) the direct linkage of the research and development (R and D) activities to the general manufacturing, financial, and marketing activities of the organisation as a whole;
(b) the framing of planned programmes of innovation in relation to the assessment of opportunities revealed by a sophisticated analysis of market situations;
(c) management which is not only effective technically, but which is market-orientated and dedicated commercially;
(d) a capability for achieving a short lead-time from the start of a new project to the marketing of the initial product; and
(e) a proper scale of production capacity and size of market in relation to the launching costs of the project.

The importance of arranging that R and D, production and marketing come under the same control, so that they constitute a single innovative activity – to ensure that new projects satisfy real market needs and opportunities; that they are matched to the production and financial capabilities of the organisation; that the commercial policy-makers are alerted to new technical means for meeting long-standing needs; and correspondingly that the scientists and technologists appreciate the commercial facts of life – has been repeatedly stressed by those experienced in technological industry.

Nevertheless, this basic requirement has been much less appreciated in the countries of Western Europe than in the United States. A recent analysis of major industrial innovations shows the effect of this clearly. Of the inventions which led to these, 10 were initiated by Britain, France and Germany, and 19 by the United States; but only 7 were converted into final product innovations by the European countries, as against 22 by the United States. One major strength of American industry has lain in its ability to carry an idea through to the final product without a break in the innovative chain. Another has been a readiness of banks and private investors to finance technological innovation. A third has been the scale and impact of Government purchasing policy. In particular, and since the end of the Second World War, the United States Government, with the enormous resources it was able to command, has supported a huge and pervasive military establishment, which has constantly demanded more and more sophisticated equipment, and which in so doing has been an extremely potent force both in raising domestic demand in general, and in generating the sense of technological awareness which characterises so large a part of the vast home market of America today.

Effective leadership by management is obviously necessary to secure the required integration of technological and trading policies. The general management – the Board of Directors in a firm – has the responsibility of setting an overall policy which takes account of all the technical, production, financial, and marketing factors; and then of ensuring that the various departments of the organisation understand the need to co-ordinate their activities within the framework of this policy, at the same time securing the full co-operation of the whole labour force in the ensuing technological changes. Strong technical leadership from management is also needed, moreover, to provide the driving force to push new projects forward against all the resistance to change which is inevitable in almost any organisation. Enthusiasm and determination are vital factors in forcing inventions through to success.

Given a decision to go ahead with a new development, a short lead-time is equally essential. One reason for this is purely economic. Innovation costs money. The shorter the time between the start of a project and its commercial fulfilment, the sooner the commercial returns. Since £100 to be received in, say, five years' time is worth far more than £100 in 30 years' time, because it could be earning interest in the meantime, long-term projects are less economically attractive than short-term ones, except when the expected pay-off from them is exceptionally high.

A scientific discovery may take many years before it finds

practical application. It follows, therefore, that the case for supporting long-term scientific research cannot be argued on economic
grounds. The justification for it is that this constitutes the fount of all
new knowledge, without which the opportunities for further technical
progress must eventually become exhausted. Laboratories carrying
out long-term basic researches which stretch techniques to the
limit are at the same time a valuable forcing ground for new technical
developments. The scale of applied research and advanced development must, however, be necessarily determined by economic considerations, since these activities may culminate in a practical
invention in something like three to seven years. A corresponding
period may be required for the engineering development which
transforms the invention into a full prototype ready for commercial
development. Minor developments may take as little as one to two
years; and newly installed production plant is normally brought into
full use in about this same time.

Several firms and countries have prospered through a policy of
avoiding long lead-times and heavy development costs by buying
other people's technological knowledge and by concentrating on the
commercial application of imported inventions and innovations.
The royalty and licence costs are usually small compared with other
commercial outlays and returns. Britain has pursued this policy less
than have some countries of similar size which because of the
devastation of war had to embark on vast programmes of industrial
reconstruction. In 1964, for example, we spent only some £50
million on royalties and licences, well below one-tenth of what we
spent on our own R and D, and less than we earned from royalties
we sold. At about this same time, France and Germany, which have
prospered greatly in recent years, each spent on foreign royalties
nearly three times as much as they earned from receipts on their own.
Japan has been particularly successful in pursuing the same policy.
Even in the United States it is common practice to buy other people's
technical knowledge. For example, 15 of the 25 major innovations
made by du Pont, a powerful science-based organisation, came from
inventions made outside the firm.

On the other hand, no country could expect to progress in
industries based on advanced technology just by buying in other
people's ideas. Such a policy succeeds when the company which
proposes to operate on a royalty basis for a particular technological
product is itself under dynamic technical and commercial direction,
and is an innovator in its own right. There have been United Kingdom companies which have gone downhill simply because they failed,
either for lack of the right kind of staff, or because management was
too feeble, to take advantage of technical information freely handed

APPENDIX 10 / 175

over by an allied American company. Equally it is known that United States subsidiaries in the United Kingdom, particularly those which are United States-managed, often do better than corresponding United Kingdom companies. Clearly vigorous commercial and scientific management is every bit as important as an idea which is being exploited, whether it be home-grown or imported. Moreover, as emphasised below, it is in the final stages of the process of innovation, in tooling up, in introducing improvements in processes, in skilful and aggressive marketing, that the biggest effort in terms of professionally trained manpower is called for. Thus, while buying royalty rights and so saving resources in the stages of R and D and invention can undoubtedly help to reduce lead-times, it cannot compensate for any lack of technological and commercial ability in the later and more costly parts of the process of innovation.

APPENDIX 11

Linear Models of Innovation

J. LANGRISH, M. GIBBONS, W. G. EVANS AND F. R. JEVONS

From *Wealth from Knowledge* (Macmillan 1972), pp. 72–6

Models are not just of academic interest; they are also important in practice since some policies seem to rest on assumptions which can be described as models. This section draws attention to some limitations of some current ones.

Most writers on innovation have either clearly stated or implicitly assumed that the innovation process consists of a linear sequence of events. These linear models of innovation can be divided into two types: those in which the start of the process is a discovery, and those in which the start of the process is some form of need.

An example for the first type, which can be called the 'discovery push' model, is provided by Blackett, who states: 'In a simplified schematic form, successful technological innovation can be envisaged as consisting of a sequence of related steps: pure science, applied science, invention, development, prototype construction, production, marketing, sales and profit. Clearly the first steps . . . cost money and only the later stages . . . make money.'[1]

An example of the second type, which can be called the 'need pull' model, is provided by Hollomon, who claims: 'The sequence – perceived need, invention, innovation (limited by political, social or economic forces) and diffusion or adaptation (determined by the organisational character and incentives of industry) – is the one most often met in the regular civilian economy.'[2]

These two types of model can each be further divided into two subdivisions, producing four models of the innovation process as follows. (The symbols DS, DT, NC and NX are used in Tables A to C to refer to these models.)

1. DISCOVERY PUSH

DS. The 'science discovers, technology applies' model
Blackett's description given above falls into this category, in which innovation is seen as the process whereby scientific discoveries are turned into commercial products. Attempts to measure time lags between scientific discoveries and their applications assume this model to be a valid description, as do comments about foreign industry being better at applying the results of Britain's science.

DT. The 'technological discovery' model
Many innovations are not clearly based on any scientific discovery
but can be described as being based on an invention or technological
discovery. For example, Pilkington's float glass process can be
regarded as being based on a technological discovery. Pippard[3] has
listed the delays between some technological discoveries and their
applications.

2. NEED PULL

NC. The 'customer need' model
Innovation can be considered as a process which starts with the
realisation of a market need. Market research or a direct request for
a new product from a customer can be the start of research and
development activity leading to successful innovation.

NX. The 'management by objective' model
Some innovations can be described in terms of the start of the
process being a need identified by the management where this need
is not a customer need. For example, the need to reduce the costs of
a manufacturing process can lead to resources being allocated to
research and development which may produce a new and cheaper
process. Another kind of example is the case of a firm producing a
new product to avoid a take-over possibility.

When the Queen's Award innovations are examined, *very few of
them fit any one of the above models* in a clear and unambiguous
manner. The reason for this is quite simple. It is extremely difficult
to describe the majority of the cases in terms of a linear sequence with
a clearly defined starting point. For example, in the case of Lytag
Ltd, a subsidiary of John Laing, the Award was given for the
development and sale of a new lightweight aggregate used in the
production of low-density concrete. If the process leading to the
manufacture for sale of this new product is to be described in linear
terms, what is the start of the process? Is it the Roman's use of
pumice as a lightweight aggregate, the manufacture of an artificial
lightweight aggregate during the First World War, the need of the
Central Electricity Generating Board to dispose of the pulverised fuel
ash used in the Lytag process, or the work carried out at the Building
Research Station which resulted in the supply of both ideas and
people to the John Laing organisation? The Lytag innovation can
also be considered to start with Maurice Laing's desire to venture
into new areas of activity, but he would not have gone ahead with
Lytag without a belief in a potential market and, from this point of

view, the Lytag innovation starts with changes in society creating a need for a material with good insulating properties.

Clearly, any of the four models given above can be made to fit the Lytag case, though the 'science discovers, technology applies' model can be brought in only in very indirect ways, such as by considering the manufacture of an aggregate from a waste product of electricity generation as an application following from the scientific discovery of electricity.

The complexity of the Lytag case is by no means unique. Innovation is a complex process involving the interaction of many factors. This complexity, however, can be simplified somewhat by restricting the process to the Award-winning firm. The question then becomes, *'What stimulated the firm into the activity that led to the successful innovation?'*

From the point of view of the Award-winning firm, it becomes possible to categorise some innovations as being clearly 'discovery push' or 'need pull'. If a sales manager realises that a product needs a particular new property and then persuades the firm to develop a product with this new property, then the innovation is of the 'need pull' type. If, on the other hand, a research department discovers a material with new properties and the firm attempts to find if the new properties have any commercial value, then it is an example of the 'discovery push' type. Similarly, if the managing director of a firm becomes fascinated with a new technological discovery and spends money on its development with no clear indication of any specific market potential, then it is also an example of the 'discovery push' type.

Even when the complex process of innovation is simplified by concentrating on what stirred the Award-winning firm into action, it is still very difficult in a large number of cases to state clearly that the innovation is of one type or another. However, if the above models are regarded as complementary rather than mutually exclusive, the innovations can sometimes be better described as a combination of two of the above models. Tables A and B show the numbers of innovations which fit the particular models. Table C is a combination of A and B with each 'dual' type scoring 2. It can be seen that, numerically, 'need pull' is more important than 'discovery push'. However . . . the larger technological changes tend to be of the 'discovery push' type.

Table A

AREA OF INNOVATION	TYPE OF INNOVATION FROM POINT OF VIEW OF FIRM						
	Not known	*DS*	*DT*	*NC*	*NX*	*Dual*	*Total*
Chemical	1	0	1	1	1	8	12
Mech. Eng.	4	0	3	13	7	13	40
Electrical	2	0	1	3	7	10	23
Craft-based	1	0	0	1	3	4	9
Total	8	0	5	18	18	35	84

Table B

	ANALYSIS OF THE 35 DUAL CASES					
	DS/DT	*DS/NX*	*DT/NX*	*DT/NC*	*NC/NX*	*Total*
Chemical	1	0	2	4	1	8
Mech. Eng.	0	0	3	9	1	13
Electrical	0	0	3	5	2	10
Craft-based	0	1	1	2	0	4
Total	1	1	9	20	4	35

Table C

	TOTAL OCCURRENCES IN DUAL AND SINGLE TYPES				
	'*Discovery push*'		'*Need pull*'		
	DS	*DT*	*NC*	*NX*	*Total*
Chemical	1	8	6	4	19
Mech. Eng.	0	15	23	11	49
Electrical	0	9	10	12	31
Craft-based	1	3	3	5	12
	2	35	42	32	111
Total	37		74		

Notes
1. P. M. S. Blackett, *Nature* vol. 219 (1968), p. 1107.
2. J. H. Hollomon, in R. A. Tybout (ed.), *Economics of Research and Development* (Ohio State University Press, Columbus 1965), p. 253.
3. A. B. Pippard, Annex D to the Swann Report, *The Flow into Employment of Scientists, Engineers and Technologists* (Cmnd 3760, HMSO, London 1968).

APPENDIX 12

Critical Science

J. R. RAVETZ

From *Scientific Knowledge and its Social Problems* (Oxford University Press 1971), pp. 423–6, 435

The process of industrialization is irreversible; and the innocence of academic science cannot be regained. The resolution of the social problems of science created by its industrialisation will depend very strongly on the particular circumstances and traditions of each field in each nation. Where morale and effective leadership can be maintained under the new conditions, we may see entire fields adjusting successfully to them, and producing work which is both worth while as science and useful as a contribution to technology. Recruits to this sort of science will see it as a career only marginally different from any other open to them; and it is not impossible for men of ability and integrity to rise to leadership in such an environment. This thoroughly industrialised science will necessarily become the major part of the scientific enterprise, sharing resources with a few high-prestige fields of 'undirected' research, and allowing some crumbs for the remnants of small-scale individual research. A frank recognition of this situation will help in the solution of the problems of decision and control. Since the criteria of assessment of quality will be heavily biased towards possible technical functions of results, they will thereby be more easily applied, and less subject to abuse, than those which are based on the imponderable 'internal' components of value.

Thus, provided that the crises in recruitment and morale do not lead to the degeneration and corruption of whole fields, we can expect emergence of a stable, thoroughly industrialised natural science, responsible to society at large through its contribution to the solution of the technical problems set by industry and the State. Scientists, and their leaders and institutions, will be 'tame': accepting their dependence and their responsibilities, they will be unlikely to engage in, or encourage, public criticisms of the policies of those institutions that support their research and employ their graduates. Such a policy of prudence is not necessarily corruption; whether it becomes so will depend on many subtle factors in the self-consciousness of this new sort of science, and the claims made to its audiences. But not all the members of any group are easily tamed, and the emergence of a 'critical science', as a self-conscious and coherent force, is one of the most significant and hopeful developments of the present period.

There have always been natural scientists concerned with the sufferings of humanity; but with very few exceptions they have faced the alternatives of doing irrelevant academic research to gain the leisure and freedom for their social campaigns, or doing applied research which could benefit humanity only if it first produced profits for their industrial employer. The results of pharmaceutical research must pass through the cash nexus of that industry before being applied, and that process may be an unsavoury one. Only in the fields related to 'social medicine' could genuine scientific research make a direct contribution to the solution of practical problems, of protecting the health and welfare of an otherwise defenceless public. Now, however, the threats to human welfare and survival made by the runaway technology of the present provide opportunities for such beneficial research in a wide range of fields; and the problems there are as difficult and challenging as any in academic science. These new problems do more than provide opportunities for scientific research with humanitarian functions. For the response to this peril is rapidly creating a new sort of science: critical science. Instead of isolated individuals sacrificing their leisure and interrupting their regular research for engagement in practical problems, we now see the emergence of scientific schools of a new sort. In them, collaborative research of the highest quality is done, as part of practical projects involving the discovery, analysis and criticism of the different sorts of damage inflicted on man and nature by runaway technology, followed by their public exposure and campaigns for their abolition. The honour of creating the first school of 'critical science' belongs to Professor Barry Commoner and his colleagues at Washington University, St Louis, together with the Committee for Environmental Information which publishes *Environment*. . . .[1]

The problem-situations which critical science investigates are not the result of deliberate attempts to poison the environment. But they result from practices whose correction will involve inconvenience and money cost; and the interests involved may be those of powerful groups of firms, or agencies of the State itself. The work of inquiry is largely futile unless it is followed up by exposure and campaigning; and hence critical science is inevitably and essentially political.[2] Its style of politics is not that of the modern mass movements or even that of 'pressure groups' representing a particular constituency with a distinct set of interests; it is more like the politics of the Enlightenment, where a small minority uses reason, argument, and a mixture of political tactics to arouse a public concern on matters of human welfare. The opponents of critical science will usually be bureaucratic institutions which try to remain faceless, pushing their tame experts, and hired advocates and image-projectors, into the line of battle; although occasionally a very distinguished man is exposed as more irresponsible than he would care to admit. . . .

The presence of an effective critical science is naturally an embarrassment to the leadership of the responsible, industrialised, tame scientific establishment. Their natural (and sincere) reaction is to accuse the critics of being negative and irresponsible; and their defensive slogan is along the lines of 'technology creates problems, which technology can solve'. This is not strictly true in all cases, since nothing will solve the problems of the children already killed or deformed by radioactive fallout or by the drug Thalidomide. Moreover, this claim carries the implication that 'technology' is an autonomous and self-correcting process. This is patent nonsense. . . . On the self-correcting tendency of technology, one might argue that no large and responsible institution would continue harmful practices once they had been recognised; but this generalisation is analogous to the traditional denial of the cruelty of slavery, along the lines that no sensible man would maltreat such valuable pieces of property. And the history of the struggles for public health and against pollution, from their inception to the present, shows that the guilty institutions and groups of people will usually fight by every means available to prevent their immediate interests being sacrificed to some impalpable public benefit. . . .

In this present period, we may find Francis Bacon speaking to us more than Descartes the metaphysician–geometer or Galileo the engineer–cosmologist. As deeply as any of his pietistic, alchemical forerunners, he felt the love of God's creation, the pity for the sufferings of man, and the striving for innocence, humility, and charity; and he recognised vanity as the deadliest of sins.[3] To this last he ascribed the evil state of the arts and sciences: 'For we copy the sin of our first parents while we suffer for it. They wished to be like God, but their posterity wish to be even greater. For we create worlds, we direct and domineer over nature, we will have it that all things *are* as in our folly we think they should be, not as seems fittest to the Divine wisdom, or as they are found to be in fact.'

The punishment for all this, as Bacon saw it, was ignorance and impotence. It might seem that the problem is different now, for we have so much scientific knowledge and merely face the task of applying it for good rather than evil. But Bacon assumed his readers to believe themselves in possession of great knowledge; and much of his writing was devoted to disabusing them of this illusion. Perhaps the daily reports of 'insufficient knowledge' of the effects of this or that aspect of the rape of the earth, and our sense of insufficient understanding of what our social and spiritual crises are all about, indicate that in spite of the magnificent edifice of genuine scientific knowledge bequeathed to us, we are only at the beginning of learning the things, and the ways, necessary for the human life.

Notes
1. The first statement of 'critical science' as distinct from ecological concern is B. Commoner, *Science and Survival* (Gollancz, London 1966).
2. The most comprehensive analysis of 'critical science' yet published is Max Nicholson, *The Environmental Revolution* (Hodder & Stoughton London 1970). He is mainly concerned with 'conservation', but his healthy approach to modern bureaucratic politics is developed in his earlier book, *The System* (Hodder & Stoughton, London 1967).
3. For a detailed interpretation of Bacon's programme for science in terms of a vision of moral and spiritual reform, see J. R. Ravetz, 'Francis Bacon and the Reform of Philosophy', in *Science, Medicine and Society in the Renaissance* (Walter Pagel Festschrift), ed. A. Debus (University of Chicago Press and Oldbourne Press, London 1972). This is an elaboration of certain themes in Benjamin Farrington's *The Philosophy of Francis Bacon* (Liverpool University Press 1964), and I am indebted to him for my first insights into this aspect of Bacon.

Index

administration 25, 31, 81, 84, 91, 114, 129, 131, 150, 155
Africa 18, 22, 126
airport 49–50
alchemy 98
American Chemical Society 14
anagrams 78
arts 52, 54, 78, 82–3, 95, 134, 182
Asia 22, 126
atom bomb 14, 21, 106, 126, 133, 170
atomic energy *see* nuclear energy
automation 131
autonomy 26, 76, 81, 99, 182

Bacon, F. 97–9, 130, 141, 142, 182–3
Bernal, J.D. 13–31, 38, 39, 114, 125–33
Blackett, P.M.S. 22, 28, 102, 127, 176
big science 27–9, 34, 84–5
books 63–4, 69–70, 71, 82–3
brain drain 92–3, 158
bread 168
broad courses 95–6, 162
Buchanan, C. 50
bureaucracy 30, 114, 154–5, 182, 183
Byatt, I.C.R. 103–7, 163–70

Cambridge 57–9
Central Advisory Council for Science and Technology 108–11, 172–5
CERN 85, 137
Chorleywood bread process 168
chromatography 45
citations 43–5
coal 17, 118
Cohen, A.V. 103–7, 163–70
collaboration 42–3, 45, 83–5, 140, 149–50, 152, 154, 181
Commoner, B. 181
communication 63, 79, 127, 156
computers 24, 30, 50, 127, 129, 131
Conant, J.B. 14
concrete 177

conflict 88, 154–6
consultants 108
contractual theory 80, 148
control 76–7, 80, 84, 146–50, 155
Copernicus, N. 55–7, 60, 61, 68, 144
Council for Scientific Policy 28, 38, 103, 167
Crick, F. 57–9, 79
critical science 144–5, 180–3
cryogenics 169
culture 48, 54, 88, 154–7, 164
curiosity-oriented research 104–8, 163–70
customer/contractor principle 101

Dainton, F.S. 92, 158
Darwin, C. 142
Davies, D.S. 93
DDT 21
determinism 30, 39
developing countries 22, 117–18, 126
discounted cash flow 105, 166, 167, 173
discovery push 102–3, 105, 111–13, 176–9
disorganisation 84–5, 149–50
dispersants 17
DNA 57–9, 68–9, 79
Dolittle Dr 113

economic benefits 21–3, 103–8, 163–70
economic growth 35, 109, 112, 116–18, 125
ecology 21, 183
education 31, 38, 42, 46, 69–70, 77, 81, 86–7, 92–6, 127, 130–1, 158–62, 163
Einstein, A. 23, 144
elitism 31, 42
Ellis, N.D. 88–91, 93, 94, 152–7
employment 92–6, 152–62
enculturation 157